JAMES CAMERON'S

# Aliens
## OF THE
# Deep

JAMES CAMERON'S

# ALIENS
## OF
## THE
# DEEP

### DR. JOSEPH MACINNIS

 **NATIONAL GEOGRAPHIC**

WASHINGTON, D.C.

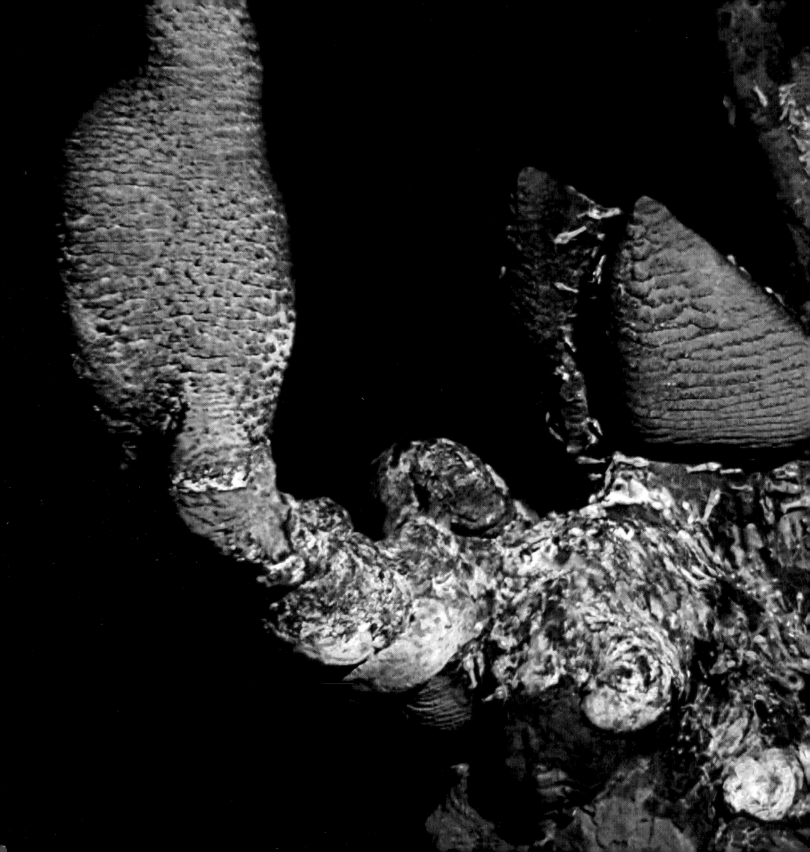

THIS PAGE: Hydrothermal vent at Snake Pit, North Atlantic
PREVIOUS PAGES: Deep-sea jellyfish *Deepstaria enigmatica*
FOLLOWING PAGES: The deck of the *Akademik Keldysh*.

BY JAMES CAMERON

When the hatch closes with a metallic "thunk," you know you're really going. All the years of dreaming, the months of planning, building, testing, and the weeks of backbreaking labor in the tropical sun fitting the subs to the new ship are behind us. This is where theory ends and reality begins. I can see my brother Dave, whom we have dubbed the Great Pumpkin because of his orange jumpsuit (and the fact that he towers over the other deck crew) barking orders, getting the winch operators and line handlers on their toes. Dave is an ex-Marine sergeant who has taken very naturally to the task of being our Surface Officer.

I hear Dave's command to the crane operator and the *Deep Rover* lifts off its cradle like a hovering helicopter. The hull is clear acrylic, five inches thick to withstand the pressures of the deep ocean, and gives an unobstructed view in every direction. We seem to fly over the side of the ship and out over the water. Time to get wet. I hear "Pay out the line, pay out the line!" over the VHS surface radio. The sea envelops us in its blue embrace. We are towed to a rendezvous with the other sub, *Deep Rover 2,* and within minutes are descending together toward the bottom of the sea half a mile beneath us.

It is a childhood fantasy of mine playing out. When I was a kid, nothing seemed more exciting, more noble and worthy, more sheer damn adventurous, than the idea of exploring an alien world. As I rode the school bus an hour each way every day, I read voraciously the best and the worst of science fiction. In those worlds of imagination, I could envision myself in a spaceship landing on another planet, the first human to gaze on its wonders, the first to return to tell the tale. Now I am really doing it.

The two of us, my pilot and I, descend through the deepening blue in our little humming bubble of air and light, and the pounds per square inch build up outside. Soon the noonday sun of the surface world has faded, leaving a deep ultramarine twilight. It is the most beautiful color I have ever seen. In some ways it is my favorite moment of the dive, suspended between worlds, saying good-bye to all you've ever known, and surrounded by infinite blue. It's a hue that not only suggests the ocean's vast scale but beckons the mind to a transcendent state—a sense of cosmic unity with

OPPOSITE: *Jim Cameron emerges from the hatch of a* Mir *sub.*

the ocean, with ancient time and the history of life, back to the first organisms.

Our expedition is an ambitious attempt to fuse film-making and scientific research. We are not only making a film about deep-ocean research, we are using the film's budget to fund that research—a first-ever experiment. We have a ship full of scientists from major oceanographic and NASA centers eager to get a look at the world they have studied only in the lab and on a theoretical level. Now they are confronting that world directly with their eyes and the heightened perception of those who project themselves physically into the heart of an unknown and hostile world. Their objective is to wrestle away a little truth, to expand the modest circle of light we call knowledge by some tiny amount.

Below us, in the darkness, is a volcanic vent site called Lost City. Its spectacular natural architecture has been built up over the eons by hydrothermal activity, for none to see, except the few bold enough to challenge the pressure and eternal blackness. Our lights will illuminate its cliffs and pinnacles, but will fail to do justice to its geological majesty.

I've made more than 60 extreme-depth sub dives, some as deep as three miles, and there is a moment before every dive—usually when I'm pulling on my woolly socks—when a thought flashes through my mind: "Is this the last time I will see daylight?" The sea offers up great gifts, but is a demanding mistress. Often great labor is rewarded with utter failure, sometimes death.

But to everyone who has seen what lies inside the heart of the ocean, the gifts are worth the hard price of admission. There is the reward of data, science, and knowledge. There are the wondrous and indelible images of creatures and natural formations that inspire our imaginations. These are the greatest gifts, because they remind us of our place in the great unfolding of the universe, simultaneously central and insignificant. As Freeman Dyson says, "Nature's imagination is so much richer than our own."

Dr. Joe MacInnis—who wrote this beautiful and insightful book about our expedition's chaotic and oft-thwarted efforts to bring back some of the sea's gifts—has spent his life exploring and understanding the ocean and the human family's relationship with it. He watched us struggle with our technology, our limitations, and the capricious elements with the patience of a man who has seen great battles won and lost in all the oceans of the world. His humor and his hard-won wisdom often inspired me at critical moments. Late one night, during a particularly dark hour, he shared with me a saying popular with the fishermen of Newfoundland. "We don't be takin' nothin' from the sea," he told me. "We sneaks up on what we wants—and wiggles it away."

On this expedition I would come to learn the wisdom of that saying. To arrogantly expect our technology to prevail in a corrosive and unforgiving environment is to invite failure or worse. Being humble before the task, vigilant about the risks, and grateful for whatever gifts are granted is the only way to enter and leave the deep ocean.

At the first meeting of all our technicians, engineers, scientists, and film crew I wrote on a big white board the three catchphrases that had become my personal mantra for the expedition:

LUCK IS NOT A FACTOR.

HOPE IS NOT A STRATEGY.

FEAR IS NOT AN OPTION.

The first two were meant to convey the critical concept that the ocean will not cut us any slack, so we must make our own luck and leave nothing to chance; that we must prepare, test, test again, and bring two backups for every piece of equipment, because we will be doing something difficult and complex in one of the harshest realms on the planet.

The third was a bit of macho posturing, a reminder that in exploring the unknown one must balance caution with boldness.

A few weeks into the expedition, we are discovering how true those statements are. We are trying mightily to make our own luck, seldom sleeping, working around the clock, and wrestling with what seem like insurmountable problems. In the end, we will fail at some of our goals and succeed at many.

But right now, poised between worlds and descending into blackness, there is only the elemental desire to see and know. It's what drives us into this alien world, and it's what will drive us as a species out into the universe, to challenge that other great blackness, which is equally mysterious and unforgiving. Right now, there is water pooling at our feet. The sub is leaking. I am hoping that we can find the leak and stop it so we don't have to abort the dive. The pilot is disassembling the penetrator-plate cover above our heads to find out where the water is coming in. Then, without warning, all the lights go out and the sub goes dead. Through the transparent crew sphere I can see *Rover 2* hovering nearby like a candle in the darkness. Now we have two problems to solve and a long way to go before we reach the seafloor.

Hopefully we can fix things and continue. We've made it this far with our shiny new cameras and lights, hoping to catch a glimpse of the deep ocean's secrets. I so desperately want to wiggle something away.

# CHAPTER ONE

# MAKING WAVES

## PREPARING FOR THE EXPEDITION

An astronaut's-eye view of our blue planet reveals its massive ocean, home to the vast majority of life on Earth.

## JULY 2003

I am standing at the stern of the *EDT Ares* watching the ship's wake unfold under the vault of the midnight sky. The wake is an expanding white ribbon streaming out behind the ship and then disappearing into the dark. In places it dances with patches of bioluminescence.

I have spent the past 50 years exploring the universe that lies below the surface of the sea. Since my first scuba dive in 1954, I have made thousands of dives in the Arctic, Atlantic, and Pacific Oceans. Each time I slip beneath the surface I ask myself the same questions: "What is down here? What does it mean? What's the best way to share this shadowy realm with the rest of the world?"

Three weeks ago I sailed out from the coast of Florida on a new expedition. We are asking the same questions, but on a scale I have never before imagined. We have two large ships and 170 talented scientists, technicians, and filmmakers, as well as four state-of-the-art research submarines. Our purpose is to explore a part of the deep ocean that few people have ever seen, a world so different from what exists on the surface

*Two miles below the surface, James Cameron directs three other sub crews from inside the pressure sphere of* Mir One.

that some have likened it to another planet. In the process we will probe several billion-year-old mysteries lying hidden in the depths. And we'll incorporate everything we see into a large-format, 3-D film.

Thousands of feet below the deck of the *Ares*, the Earth is breaking apart. Along the Mid-Atlantic Ridge, a mountain chain that snakes the length of the Atlantic from Greenland to Antarctica, the crust of the planet is splitting along its tectonic plates. At the plate boundaries hot magma wells up from beneath the crust, creating a central rift valley and countless openings called hydrothermal vents. In these superheated vent environments, far from the reach of the sun, scientists have discovered bizarre communities of life. They thrive in a place where there is very little life-giving photosynthesis, temperatures reach more than 500°F, toxic chemicals

OPPOSITE: *Loaded with tons of equipment and 40 people, the* EDT Ares *maneuvers into position above the Mid-Atlantic Ridge.*

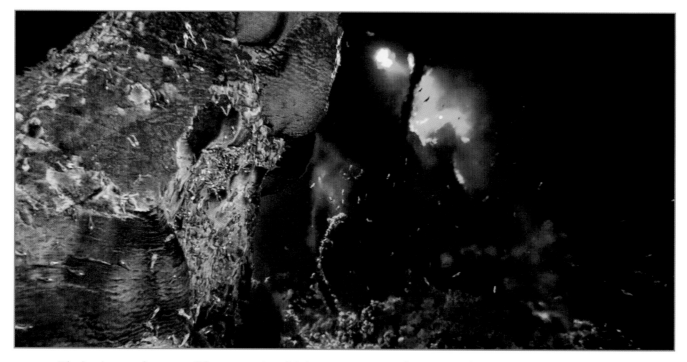

*Rimicaris exoculata, a prolific vent species of shrimp, swarm around a steaming hydrothermal vent at Snake Pit.*

spew forth, and the intense pressure is lethal to humans. Even more astonishing, the vent organisms include some of the most primitive life forms on the planet. They offer tantalizing clues about the rise of life on Earth, and may hold the key to finding life on other planets.

Only a handful of research subs can dive to these depths and explore these unfamiliar landscapes. Everyone on these two ships is awash in the excitement of seeing this alien world—Earth's final frontier.

The 247-foot-long *Ares* has a wide main deck and a high white superstructure. She is a new ship that can accommodate 50 people working on deep-sea survey and recovery projects. The *Ares* has been chartered for this expedition because she has the crew and the machinery to put heavy objects into the ocean and a combination of computers and thrusters that allow her to hold a fixed position—in strong winds and currents—without using her anchors.

We are more than a hundred miles south of the Azores, steaming toward our next dive site. Two miles ahead are the lights of a second vessel, the *Akademik Keldysh.* Operated by the Russian Academy of Sciences, *Keldysh* is the world's biggest research ship. Within her 450-foot length are 130 people, 17 laboratories, a sauna, and a swimming pool. Secured to her main deck is a pair of 20-million-dollar minisubs called *Mir One* and *Mir Two.* Slightly larger than a delivery van, each *Mir* can take three people to a depth of 20,000 feet and work at that depth for 15 hours. Several years ago, one of them carried me down to the rusting decks of the *Titanic.*

A few steps away in the center of *Ares's* main deck, two smaller subs, *Deep Rover*s *1* and *2,* are berthed. As sleek as a pair of jet fighters, each *Rover* is equipped with a transparent Plexiglas sphere that gives their two-person crew a panoramic view of the inside of the ocean. A titanium pressure housing mounted on the bow of *Deep Rover 1* holds a 3-D high-definition (3-D-HD) camera. One of only four such cameras in the world, it is worth a quarter of a million dollars.

For the past week we have been using the *Mir*s and the *Rover*s to explore and film hydrothermal vent fields in this section of the Atlantic. Many vents lie in water more than two miles deep, but we have been visiting some much closer to the surface.

A tall man in a dark blue T-shirt steps up to the railing beside me. Jim Cameron leans forward, places his strong forearms on the cool steel, and gazes at the black continuum of sea and space. Riding high in the heavens to the east is the full moon with Mars above its shoulder. Light from the moon falls upon the sea like a magic, unearthly dawn.

He responds quietly to my greeting. As the leader of one of the most complex deep-sea expeditions ever conducted, Cameron has a lot on his mind. How good is the footage he has just shot on the volcanic vent fields almost 3,000 feet below the surface? Can the sub crew resolve the electrical problems haunting both *Deep Rover*s?

My own thoughts are on two glowing objects—the moon and Mars. In a few days the red planet will be as close to the Earth as it has been for 60,000 years. Taking advantage of this proximity, NASA recently launched two robotic vehicles, *Spirit* and *Opportunity*, that are currently heading toward its dusty surface. When they land, they will begin searching for signs of ancient water in the nearby rocks. Recent data from an earlier orbiting spacecraft hint that Mars's northern lowlands may once have been covered with an ocean. An ancient Martian sea would be a tantalizing sign that life may once have thrived on the red planet.

Most people know Jim Cameron as the Hollywood director who made *Aliens, The Abyss, True Lies,* and *Titanic.* But he's also a serious undersea explorer. For him, exploration is the elixir of life. Since their first dives to the *Titanic* in 1995, Cameron and his team have been using research subs and mini-robots to probe the ocean's great secrets

Cameron the ocean explorer has always had a fierce interest in the science and technology of outer space. He reads journals, scans the Internet, and talks to senior scientists at NASA, looking for facts of compelling interest. He knows that Europa, one of Jupiter's moons, is covered with ice several miles thick and that the hypothetical ocean under the ice could be at least as big as all the oceans on Earth. He is aware that many scientists believe that if life exists somewhere else in the universe, Europa is a

strong candidate. Whatever else happens, the NASA scientists aboard our ships will come away with a better understanding of the challenges of exploring a lethal, high-pressure environment.

What Cameron wants to do with this film is combine the facts he knows about Earth's ocean and outer space into a believable story.

Like all his stories, it has to have an authentic context. That means going to the black bottom of two oceans and creating a moon-bright landscape for a firsthand look at what is down there: the smoking chimneys that roar, the legion of animals that live under deadly plumes of minerals, the sulfur-eating microbes that underpin the entire ecosystem.

This two-month expedition is the latest expression of Cameron's passion for the deep ocean. It began when he was a teenager growing up in Niagara Falls, Ontario. In the mid-1960s he started reading everything he could about the pioneering undersea work of Jacques-Yves Cousteau, Edwin Link, and the U.S. Navy. He watched the *Sea Hunt* television series, took diving lessons, and became a certified scuba diver. By his early 20s, he was already thinking about how to incorporate advanced undersea technology into a feature film.

But Cameron's ambitions went beyond making films *about* the ocean. A true explorer, he wanted to make films *under* the ocean. This is why he started the production of his 200-million-dollar movie *Titanic* by chartering the research vessel *Keldysh* and the *Mir* subs to make 12 dives to explore the world's most famous shipwreck. He and his team went on to make a 90-minute documentary on the *Bismarck*, and then returned to the *Titanic* to take audiences inside the wreck of the infamous ship for a large-format film called *Ghosts of the Abyss*.

Cameron is taking a similar approach to this summer's expedition. He's invited a team of ten scientists to participate in dives at hydrothermal vent sites in the Atlantic and Pacific. Among them are experts in microbiology, marine biology, planetary science, and astrobiology.

The scientists on this leg of the expedition are young, burning with enthusiasm, and eager to share their knowledge. Planetary scientist Kevin Hand, 29, is short and soft-spoken, with three degrees so far. As a graduate student at Stanford University, Hand is studying the possibility of life on Europa. His dives in the *Mir*s and *Deep Rover*s will give him his first view of the seafloor and help him make connections between Earth's deep-sea vents and the life-support processes on other planetary bodies.

I spoke to tall, dark-haired Kelly Snook as she took her first look at the *Deep Rover* subs. Snook, a planetary scientist at NASA's Johnson Space Center in Houston, develops programs that use the Earth and the moon to prepare for the eventual exploration of Mars. At her workstation on the *Keldysh*, she documents the scientific dives, records the geological and biological samples brought up from the seafloor, and discusses the results via satellite phone with other scientists back on land. One of her objectives is to "learn how to explore space without leaving the Earth."

Cameron steps back from the railing into the sodium glow of the ship's lights. "It's a good science team," he says. "They're hot-blooded about their specialties and they're giving us the latest thinking on the fundamental processes of extreme life."

Cameron is giving them a chance to explore a world most of them have never seen.

He gazes out on the shimmering surface of the sea and repeats a favorite catchphrase. "Exploration is like a muscle," he says. "To maintain its strength and coordination, you have to keep exercising it."

# EXTREME MACHINES

Three "extreme machines" assisted Jim Cameron and his crew in the filming of *Aliens of the Deep.* The two 18-ton, orange-and-white *Mir* subs (pictured on pages 7 and 39) have a nickel-steel pressure hull that can carry three people—the pilot and two observers—to a depth of 20,000 feet. Three small view ports in the forward part of the hull allow direct observation. Operated in tandem by a Russian crew, the *Mirs* carry a large bank of lights, travel long distances underwater, and remain at depth underwater for up to 15 hours. Both *Mirs* carry a payload of 650 pounds and can support life for 250 man-hours.

The 20-million-dollar *Mir* subs were built by the Russian Academy of Sciences for deep-sea research. They have a pair of mechanical arms, collection baskets, core drills, water sampling bottles, and plankton nets. Their scientific instrumentation includes oxygen, temperature, and heat-flux probes.

Each *Mir* has a 15-horsepower stern thruster and

two 5-horsepower side thrusters operated by the pilot's joystick. The teardrop shape of the outer hull and the moveable wing on each tail fin allows them to hover in the water and turn slowly on their own axes.

The six-ton *Deep Rover*s can carry two people—the pilot and an observer—to a depth of 3,280 feet. Their most notable feature is a five-inch-thick transparent acrylic sphere that gives the crew a panoramic view. To squeeze into their side-by-side seats, the crew slips through an 18.5-inch hatch at the bottom of the sphere. Mounted above and below the seats are internal lights and life-support, communication, navigation, and safety systems. Like a helicopter, a *Rover* is maneuvered through a multipurpose joystick that controls four variable speed thrusters.

To film places inaccessible to the subs, a nimble little robot named *Jake* was deployed from the bow of *Mir Two*. About the size of a microwave oven, *Jake* and his twin *Elwood* were originally built by Jim Cameron's brother Mike, to "fly" through the interior of *Titanic*.

*Ready for another dive, a* Deep Rover *sub slips beneath the surface. The bubble-shaped sphere gives the crew a 320° view of the surrounding ocean.*

In 2001 these ROVs (remotely operated vehicles) took astonishing pictures of the great ship's rusting stairwells, collapsed corridors, and staterooms. For *Aliens of the Deep*, Mike Cameron crouched inside the pressure sphere of *Mir Two* controlling *Jake* through a hair-thin fiber-optic cable.

To film certain segments of *Aliens of the Deep*, the operators of the three types of deep-sea vehicles worked out a complex choreography of timing and movement. The dependable *Mir* subs provided the illumination for the underwater scenes, while the cameramen took advantage of the visibility afforded by the *Deep Rover* subs. A camera inside *Jake* recorded additional footage.

"Vent organisms exist far outside the familiar domain of human physiology," says Jim Cameron. "The extreme machines allowed us to film the extremophiles."

OPPOSITE: *Mike Cameron works on the remotely operated vehicle, or ROV, nicknamed* Jake. *Developed for exploring the interior of* Titanic, *it is controlled from a* Mir *sub.*

*Kevin Hand, Kelly Snook, and Tori Hoehler process rock samples in the shipboard lab for later study.*

Human exploration has expanded greatly in the past two generations. When I began working under the ocean as a physician-scientist in the 1960s, John Glenn and the other "original seven" astronauts were being fired into space and making their first orbits around the Earth. Then came the Apollo program and the first men walked on the moon. At the same time, ocean explorers were diving to great depths and learning how to live for weeks under the ocean in undersea stations. In both realms it was a golden age of exploration.

Today, the money for deep-sea research is much harder to find, and the human exploration of outer space is restricted to an expensive, low-orbit space station. As might be expected, a recently released report from the National Science Foundation confirms the disheartening news that the numbers of young Americans seeking careers in science are in serious decline.

This expedition is Jim Cameron's way of pushing in the other direction. He is treating parts of the project like a simulated space mission. He is producing a film that will show that science and exploration can be fascinating, productive, and fun.

After leading three major deep-sea expeditions, Cameron knows that the ocean is a vast, complex place with lots of surprises. "Exploring the ocean is ten times as hard as you think it is," he says, "and you start off thinking it's really hard." He has learned through bitter experience that exploring the ocean's depths and filming what's in them is consuming, taxing work. Because of this, everything he does is grounded in scientific principles and heavy-duty research.

Two years ago, after Cameron and his team filmed the *Bismarck* in 16,000 feet of water, they steamed to these remote waters south of the Azores and aimed their 3-D underwater cameras at five hydrothermal vent fields called Lucky Strike, Lost City, Broken Spur, Snake Pit, and TAG site. The spectacular images confirmed that an exciting undersea story was waiting to be told.

Five months ago, Cameron invited a small group of us to a two-day meeting at his ranch near Santa

Barbara, California. On a sunny day in mid-February, I joined eight men and one woman around a big table next to the great fieldstone fireplace in the ranch's main building. The table was covered with computers and notepads. Among those present were Cameron's brother Mike, an aerospace engineer who designed the two mini-robots that probed the interior of *Bismarck* and *Titanic*, and his brother John David, or "J.D.," a computer-media specialist. Sitting next to them were Ed Marsh, the creative producer and senior editor at Cameron's Earthship Productions; Andrew Wight, Earthship's producer; Vince Pace, director of photography; and Patrick Lahey, the leader of the *Deep Rover* sub team.

During his 12 dives to the *Titanic* in 1995, Jim Cameron learned that a deep-sea expedition consists of two kinds of technical assets: reliable and unreliable. The Russian ship the *Akademik Keldysh* and its two *Mir* subs are as dependable as a Swiss train schedule. With hundreds of successful dives behind them, the pilots of the *Mirs* could put the scientists and their Earthship 3-D cameras at any depth down to 20,000 feet.

But Cameron was less sure about the *Deep Rovers*

that he and his friend Mike McDowell had recently purchased in France. They were eight years old, and their wiring and electronic components were suspect. They would have to be taken apart, completely rebuilt, and certified according to the exacting standards of the American Bureau of Shipping. New lights would have to be added inside and outside their pressure hulls. *Deep Rover 1* also had to be modified to carry a 3-D-HD camera in a titanium pressure housing.

For two days we talked about submarines and cameras and robots. We discussed deadlines and dollars, logistics, seaworthiness, schedules, and safety. Because we were planning to use four subs and put more than two dozen people deep into the ocean—some for the first time—safety was uppermost in our thoughts.

Two weeks before our meeting at the ranch, the space shuttle *Columbia* had disintegrated as it reentered the atmosphere at 12,500 miles an hour. The breakup of the shuttle started just a few miles north of where we were sitting. Images of the flame and the sun-bright streamers of white vapor, repeated over and over on television, were engraved deep within our minds.

At the end of the two-day meeting, I drove down to the wide, curving beach below the ranch and stood at the Pacific shore. The first people to explore the world's largest ocean were the Polynesians 3,000 years ago. They used stick maps, star clusters, sea currents, and migratory bird patterns to find their way. We were going to go below the surface of the sea and use transponders, sector-scanning sonar, and global positioning satellites to find our way. The technology was different, but one element remained the same: the peril of the sea.

The ocean is so huge it is easy to get lost or even disappear. There are so many currents, so much darkness, and so many directions to go. Any technology, no matter how sophisticated, is as prone to error as the people who use it. Thirty-four years ago, in cold waters not far from this beach, a young man named Berry Cannon was killed on a 10-million-dollar U.S Navy project called *SeaLab 3*. He died because the people in charge of the project were overworked and overconfident. How do I know this? Because I was there the day after it happened.

OPPOSITE: *Like an underwater helicopter, the nimble* Deep Rover *subs maneuver easily in any direction.*

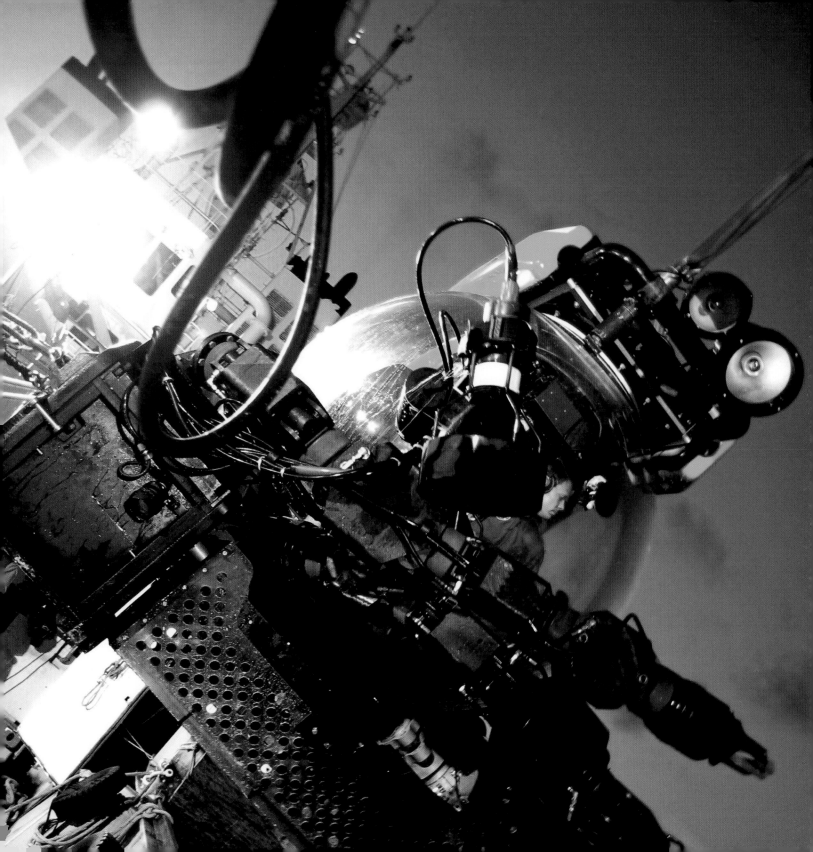

Two weeks after the meeting, a pair of 18-wheel semi-trailers rolled through the gates of the ranch and came to a stop inside a steel-frame building below the main house. Patrick Lahey, the sub crew, and six ranch hands unloaded the two *Deep Rover*s onto the building's concrete floor and began stripping them down to their bare hulls. In the process, they uncovered a gumbo soup of technical problems.

The subs' main batteries and their protective pods had to be replaced. Most of the electrical cables, cable connectors, and junction boxes were corroded. The penetrator plates—round pieces of polished metal where electrical cables and gas pipes ran from the exterior to the interior of the pressure hulls—were pitted with rust. The routine service-and-maintenance job they had envisioned turned into a costly and time-consuming major overhaul.

One of the objectives of the *Deep Rover* team was to mount heavy-duty HMI movie lights and the 3-D-HD camera with its pressure housing and pan-and-tilt unit on the bow of *Deep Rover 1*. This meant

*Guided by star clusters and migratory bird patterns, ancient Polynesians were among the first explorers to navigate the open ocean.*

weaving thick bundles of wires from the lights and cameras into the junction boxes that also controlled the propulsion, communication, and hydraulic systems.

At first they worked 16-hour days and six-day weeks. As the problems mounted, they shifted to 20-hour days and seven-day weeks. "It was a much harder job than we thought," said Lahey. "As soon as we solved a problem, another more serious one jumped into its place."

During this time, in his engineering facility 80 miles away in Valencia, California, Mike Cameron and his two-man staff were working at flank speed, trying to get the two robots ready. Mike Cameron, 47, led the team that built the pressure-proof housing and dome port for the 35mm camera that filmed the underwater scenes for *Titanic*.

Each morning before dawn, they parked their pickup trucks next to a modest cream-colored building marked with a small sign that said *Dark Matter LLC*. Mike Cameron's company specializes in the design and development of systems for deep-ocean exploration and filmmaking. At their workbenches and large-screen computers they focused on inspecting, testing, and packing the thousands of items that would have to be shipped to the *Akademik Keldysh* in St. John's, Newfoundland.

The identical robots, called *Jake* and *Elwood*, had proved themselves on the 2001 expedition to the *Titanic* and become the mechanical stars of *Ghosts of the Abyss*. They had also taken astonishing footage of the interior of the German battleship *Bismarck*. The 2,700 precision parts jammed into each of their interiors included high-energy batteries, a small computer, a television camera, LED lights, low-rpm thrusters, and dynamic buoyancy control. Their technical heart was a slim white spool that automatically spun out 2,500 feet of fiber-optic cable. Originally the highly advanced deep-sea robots were designed to slip through one of the *Titanic's* cabin windows, explore her collapsed interior, and send high-resolution video images back to their mother subs.

The multiple precision parts and complex internal architecture of the 'bots made them difficult to prepare, test, and operate. "To give you just one example," says Mike Cameron, "the fiber-optic threads that carry the data into and out of the 'bots are two-thirds the diameter of a human hair. The tips of the threads have to be hand-polished to remove any micro-scratches. It's a painfully slow, don't-make-any-mistakes-or-you-kill-the-patient process."

As the time for readying *Jake* and *Elwood* ran out, Mike Cameron made the decision to use just one robot. He scavenged parts from *Elwood* and integrated them into *Jake*. Then his team packed hundreds of spare parts, tools, and a launch and recovery enclosure into 30 large cases and shipping containers. Late one hot California night they lifted them onto a long-haul van that backed out of the industrial loading area and disappeared down the street. Like everyone else trying to meet the departure deadline, Mike Cameron knew that their task was unfinished and they would have to work overtime on the *Keldysh* while she was under way to the first dive site.

Ever since the meeting at Cameron's ranch, Andrew Wight had been trying to figure out how to pay for what was rapidly becoming one of the most expensive documentary films ever made. In the 1980s, Wight began making documentary films in

Australia about underwater exploration. As producer, Wight is responsible for helping raise the funds for the project and ensuring they are effectively spent. Late in June, he was informed that the film contract was ready for his signature. The timing couldn't have been better. The *Keldysh* was already steaming across the Atlantic toward St. John's, and the *Ares* had left her home port in Cyprus for Fort Lauderdale, Florida. At the same time, two flatbed truck convoys loaded with tons of diving and filming equipment were driving across the California desert heading for the two East Coast seaports.

At a desert junction east of Palm Springs, one of the truck drivers of the southern convoy stopped his 18-wheel vehicle and got out for a routine inspection of the strange, tarpaulin-covered cargo he'd picked up at Cameron's ranch. As he walked around the back of his 40-foot truck,

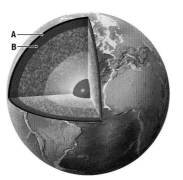

*Seawater seeping down through cracks in the Earth's crust is heated to extremely high temperatures by magma in the upper mantle (B) until it is forced back through the crust (A), at hydrothermal vents. The mantle is made up of minerals rich in the elements iron, magnesium, silicon, and oxygen. Like ice on a pond, Earth's crust, or lithosphere, floats over its molten interior. The crust is cracked in many places, forming large slabs of rock, or plates, thousands of miles across and several miles thick. As the plates move against each other they push up mountains, generate earthquakes, and spawn volcanoes.*

he noticed a pool of clear liquid oozing out of a container next to one of the *Deep Rover* subs. It had no smell, but it sure as hell looked suspicious and on this tar-hot, post-9/11 day in the desert he wasn't taking any chances. He called the local police, who, along with local firefighters, rushed to the scene.

A few hours later when Mike Cameron arrived, he found a dozen men standing around the truck wearing orange head-protecting hazmat suits. They were staring down at the pool of mysterious fluid expanding in the heat. Mike tried to tell them it was a nontoxic, mineral-based oil used to pressure-compensate the sub's batteries. The 55-gallon drum holding it must have punctured, he explained, because the truck had been packed too tight. The men in the hazmat suits kept looking at him as if he was a specimen on a swab. Being a

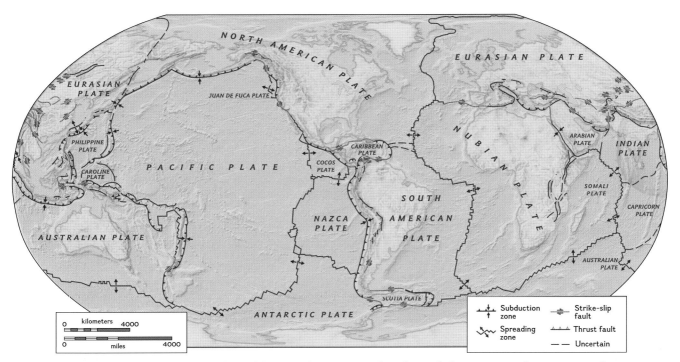

*Hydrothermal vents form near the edges of the spreading tectonic plates beneath the ocean. Moving at rates as slow as a growing fingernail, the 16 major plates that constitute Earth's crust are in constant motion.*

Cameron, Mike knew that every ambiguous story needs a clear resolution, so he dipped his finger in the liquid, stuck his finger in his mouth, and smiled. As soon as he did this, the hazmat suits started coming off and the men wearing them broke into grins. In less time than it takes to uncap a bottle of Coors Light, the lonely stretch of cactus-fringed road was deserted.

A week later, on a fiercely hot July night, the *Ares* was tied up to a long concrete pier in Fort Lauderdale, Florida. Every few minutes another jet aircraft from the international airport roared out over the Gulf Stream. On the main deck, amid the sounds of hammers and chippers and acetylene torches causing showers of sparks, men were welding a narrow-gauge railway track to the center of the *Ares's*

*Destined for a vent dive, a* Mir *sub is launched over the side of the Russian research vessel* Akademik Keldysh.
*The launch and recovery of the subs was one of the most difficult aspects of the expedition.*

main deck. At one end of the 72-foot-long track were the two gleaming *Deep Rovers,* tied down to their trolleys. At the other end, mounted over the stern like a giant wishbone, was a two-story, industrial A-frame that would lift the subs off the deck and lower them into the ocean.

Launching and recovering a sub is the most difficult part of any dive. As soon as it is lifted off its cradle, the sub starts to swing, slowly at first, but if the ship is rolling in big waves, the sub gains speed and momentum. If the launch crew loses control, they are confronted with a six-ton wrecking ball trying to destroy their ship.

During the past 20 years, the deck crews of French and American scientific institutions have learned how to launch their research subs without harming the ship, the sub, or the people inside. They use a big A-frame to lift the sub over the stern, the most stable part of the ship. They attach at least two control lines to each side of the sub and employ winches, capstans, and a strong-armed deck crew to dampen any sideways motion.

I looked down the deck at the seven men in hard hats and sweat-stained coveralls swarming over the *Deep Rover*s. Tomorrow, when the tracks were finally laid down, they would start rigging a network of nylon lines around capstans and bollards. They would winch *Deep Rover 2* on its trolley along the narrow-gauge track until it was directly under the A-frame. One of them would climb on top of *Deep Rover 2*, reach up, and hook a lift line from the center of the A-frame into a ring on top of the sub. Control lines would be fastened to both sides of the sub and then drawn tight. The A-frame operator would lift the sub off its trolley and swing it slowly backward until it was suspended over the stern of the ship. Then, inside a web of taut lines, he would lower it gently into the water.

Normally, the crane operator, winch men, and line handlers need weeks of practice to coalesce into a smoothly functioning launch and recovery team. But time was a luxury the men on the *Ares* did not have.

They were behind schedule and had only a few days to learn how to lift and lower the subs.

During a break in the welding, Jim Cameron and Mike McDowell walked across the deck and examined the 3-D-HD camera mounted on the bow of *Rover 1*. Cameron leaned into a handheld intercom and began talking to Vince Pace, a powerfully muscled man in his late 40s, sitting inside the sub. Pace and his team had built the camera and fitted it inside its gleaming, pressure-proof housing.

The camera and three others just like it, were the technical centerpiece of Cameron's expedition. The complex, precision instrument made two identical, high-definition video images and transmitted them through a bundle of wires and processors to a pair of high-definition recorders. Despite all the money and man-hours lavished on the camera mounted on *Rover 1*, despite all the redundancies that protected it against the salt air, the summer's heat, and the crushing weight of the ocean's depths, the camera was not working.

Vince Pace, staring at the camera's video monitor, was troubleshooting the problem. Sweat dampened his forehead and ran down his arms. As he spoke to Jim Cameron, he selected his words carefully. He knew the pressure Cameron was under to

# SHORES WITHOUT SUNLIGHT

In February 1977, two geologists, John Edmond of MIT and Jack Corliss of Oregon State, climbed in the research sub *Alvin*, closed the hatch, and descended to the top of the Galápagos Ridge in the eastern Pacific Ocean. Guided by a geological and thermal map that suggested the possibility of hot springs, they were looking for direct evidence of the volcanism that had created the ridge.

Like most marine scientists of that time, they thought that life in the deep sea was sparse and grew slowly in the pitch-black, near-freezing depths. This long-held premise seemed to have been confirmed nine years earlier when the same sub carrying them down to the ridge sank in 5,000 feet of water when her lift cable snapped. The men inside *Alvin* scrambled to safety, leaving their bologna sandwiches behind. When the sub was recovered eleven months later, the sandwiches showed no sign of bacterial degradation. The bread and meat tasted salty, but were good enough to eat. The conclusion was that life unfolded in the near-freezing waters at a slower pace than in a refrigerator.

When the geologists arrived at 8,000 feet below the surface, they looked out the porthole at the bleak basaltic terrain typical of a ridge axis. They asked pilot Jack Donnelly to stop and pick up some rocks. And then, in John Edmond's words: "We came upon a fabulous scene.…An oasis…reefs of mussels and fields of giant clams bathed in shimmering water, along with crabs, anemones and large pink fish."

The geologists were looking at a landscape overgrown with creatures no human had ever seen. Dozens of foot-long clams stood out chalk white against the black basalt. Limpets, brittle stars, and anemones littered the rocky seabed. Within a field of brown mussels floated fish the color of bubblegum. Blind white crabs scuttled everywhere. They saw pale orange balls with long filaments that looked like seeded dandelions. In some places the water was milky blue with bacteria. Later they found themselves hovering over a thicket of blood-red worms protruding out of hard, white tubes. On the other side of the research sub's porthole was a new chain of life—one of the most important biological discoveries of the century.

This strange landscape was a community of life surrounding a hydrothermal vent, a place where warm water rises from deep within the Earth and bursts through the seafloor. In the years since the discovery of hydrothermal vents, marine scientists from universities and oceanographic institutions have made thousands of dives to the Galápagos and other similar sites in the Atlantic and Pacific. They have discovered more than 500 new species of animals. During each dive they ask different versions of the same question: What biological and geological processes allow this miracle to exist? Part of the answer lies in the submicroscopic world of biochemistry.

Life on the Earth's surface exists because of photosynthesis, the chemical process by which the cells of green plants use sunlight to convert carbon dioxide into complex carbohydrates. These carbohydrates are the basis of the photosynthetic food chain. From microbes to mammals, we are all solar powered.

Life exists because of a delicate dance of electrons.

OPPOSITE: *Unique to hydrothermal vent communities, red-topped* Riftia, *more commonly known as tube worms, thrive at vent sites in the Pacific Ocean.*

In photosynthesis, a photon of light lifts an electron in chlorophyll— the plant cell's sunlight-catching molecule—to a higher energy level. As the electron cascades to a lower energy, it drives enzymatic reactions that make ATP, the universal energy of all living things on Earth. At the bottom of the cascade, the electron is accepted by another molecule, NADP, and reduces carbon dioxide to carbohydrate. The plant cell restocks itself with these critical electrons by splitting water molecules and releasing free oxygen into the ocean or atmosphere. So in addition to providing the food we eat, photosynthesis is responsible for every breath we take.

But the organisms at the vents don't have access to the life-giving rays of the sun. Scientists soon discovered that, deep in the ocean, a process of chemical synthesis, or chemosynthesis, takes place inside the cells of certain bacteria living in places containing water and hydrogen sulfide. At the vents, seawater is the source of oxygen (though this oxygen is

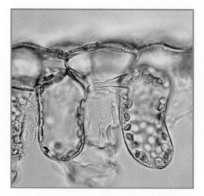

*Key to the life cycle on the Earth's surface, plant cell chloroplasts (above) convert sunlight to energy through photosynthesis. At the vents, far beyond the reach of the sun, organisms convert chemicals to energy through chemosynthesis.*

ultimately derived from photosynthesis by the surface biosphere). The hydrogen sulfide comes from cold seawater seeping down through cracks in the ocean floor, being heated by magma chambers, and boiling back up through the rocks.

In chemosynthesis, both oxygen and NADP are used as electron acceptors. In addition, the source of electrons and energy is hydrogen sulfide, a compound that is already reduced. Sulfur-eating bacteria oxidize hydrogen sulfide, using its electrons and latent energy to reduce carbon dioxide.

Chemosynthesis—using chemical energy instead of light energy—is not unique to the life in the deep ocean. All forms of life except plants, algae, and photosynthetic bacteria are chemosynthetic. What makes chemosynthesis at the vents so biologically unusual is that it takes place at the conjunction of searing heat, intense cold, ejecting chemicals, and crushing pressures.

As the years passed, Jack Corliss looked at the

data pouring in from all the vent studies and began to wonder if life on Earth had originated under similar conditions. With two of his colleagues, he proposed that life was kick-started four billion years ago, when the seafloor was more thermally active than it is today. He theorized that the water issuing from the ancient rocks was hot enough to crack the molecular bonds of the minerals and release carbon and carbon compounds such as methane into solution. The simple organic compounds adhered to the interstices of the rocks, where they formed more complex organic molecules including nucleic and amino acids. Eventually, these "precells" and "precell communities" evolved into free-living organisms.

Corliss's daring proposal initiated a fierce scientific debate that is still raging. However, many scientists are in general agreement over his premise. Recently, Cindy Lee Van Dover, a former *Alvin* pilot and now a professor of biology at the College of William and Mary, reviewed the material written about the subject and wrote: "Where is this original homestead of life? Where hot volcanic exhalations clash with circulating hydrothermal water flow …a place deep down where a pyrite-forming autocatalyst once gave and is still giving birth to life."

In the early nineties, Thomas Gold, a theorist at Cornell University, made a leap of logic that moved the debate in a different direction. He suggested that microbes—living on chemicals and heat from the Earth's core—are widespread throughout the upper three to six miles of the Earth's crust. He calculated that the mass of this deep life was approximately 200 trillion tons. "This would be…more than the existing surface flora and fauna," he wrote. "At the very least…comparable to all the living mass on the surface." He also speculated that subterranean life might exist on other planets and moons, hidden beneath their surface, powered by unseen geological forces. Our solar system, he noted, might have at least ten planetary bodies with deep biospheres supporting alien microbes.

It was a delicious thought, the kind that inspires rigorous thinking and raw speculation. Life might be thriving in faraway places like Mars, the moons of Jupiter, and other places with the right ratio of water, energy, and heat-responsive molecules. Within the scientific community, the idea grew and spread with the vigor of the vent sites that spawned it.

make this complex expedition succeed. A veteran of the expeditions to *Titanic* and the *Bismarck*, he was also familiar with Cameron's highly combustible temper. He took little comfort in the knowledge that on this sticky, steamy night, it was not only the camera that wasn't working. If the sub he was sitting in were a patient in a hospital, it would be headed for the intensive care unit.

In St. John's, Newfoundland, Anatoly Sagalevitch glanced out the window of his big cabin on board the *Akademik Keldysh* as he talked to two visitors. A short, impatient man with thick glasses and a ring of white hair, Sagalevitch was anxious for the two semi-trailers on the dock below the ship to finish unloading their cargo. The scientists and film crew and their mountain of crates and containers were on board. So were Mike Cameron's team and their mini-robot. But boxes and other large objects were still coming off the trucks. The captain was impatient to cast off the lines and take the *Keldysh* out into the Atlantic.

An electronics engineer by training, Sagalevitch assisted in the design and construction of the two *Mir* research subs in 1987. Since then, he and his eight-man *Mir* team have made hundreds of dives in the Atlantic, Pacific, and Indian Oceans. Today, as head of the Manned Submersibles Laboratory at the Shirshov Institute of Oceanology in Moscow, he is responsible for planning and coordinating every dive made by the *Mir*s.

When the Soviet Union collapsed in 1991, the Shirshov Institute of the Russian Academy of Sciences had to turn to its international friends to help pay for the ship and the subs. One of those friends was Jim Cameron.

"Cameron chartered the *Mir*s for his first dives to the *Titanic* because they have twice the battery supply of any other deep-sea sub," Sagalevitch says. "This meant he could add thousands of watts of high-intensity lights. In addition, he wanted to dive the two subs together so he could put one in front for scale and lighting while the other filmed it from behind. This is how he got some of those great dramatic shots in the opening scenes of his movie."

On the wall opposite his desk were framed, full-color posters from Cameron's three undersea movies. Sagalevitch looked out the window and saw the trucks driving away from the pier. Without smiling, he

*Measuring more than 375 miles across, the Olympus Mons volcano is one of Mars's most prominent features. The search for signs of ancient water—an indicator of the possibility of life—is a major focus of exploration on the red planet.*

walked slowly across the room, lifted a bottle, and poured ice-cold vodka into three small glasses.

The stern of the *Ares* is one of the few places on the ship where you can go for a moment of solitude. Whenever there is a break in the work I walk past the subs until I am standing next to the chest-high railing. It's a good place to reflect on the immutable ocean we are steaming across and the team that is challenging her depths. Although they would blush at the word, they are pioneers on Earth's final physical frontier. They are driven by their love of risk and the awareness that their labor adds something to the store of human knowledge. When the fury of their work is upon them, it means running up and down stairwells to fetch tools or cameras, and meals snatched quickly at whatever time of day or night they can steal away from their tasks. It means sleepless nights, taut nerves, and exhaustion in the morning.

Whenever they come out on deck, or ride between the ships in inflatable boats, they keep a watchful eye on the ocean. They have been told that she is life's cradle and home to most of Earth's living things. But they know her as moody, ominous, and inscrutable. For them, the ocean is the enemy that never sleeps. The oldest, largest physical feature on the planet doesn't give a damn about who you are or where you came from. She only cares about your capability, confidence, and commitment. Make one small mistake, commit one act of ignorance, indifference, or carelessness, and you are certain of a quick passage into the next life.

It is high noon and the sun's reflection hangs dead center over the ocean, a white, trembling promise as old as time. Sunlight—the beam of life—

travels more than 90 million miles from its thermonuclear origins and penetrates the atmosphere and the upper layers of the ocean. At a depth of a thousand feet it is almost extinguished, and the water below it is a cosmic black, yet the deep ocean is a place of immeasurable variety, unending fecundity, and incessant movement.

Most of the surface of the earth—197 million square miles—is covered with seawater more than a mile deep. This means that the deep ocean is by far the planet's dominant habitat. Most of this water is forever dark, illuminated only by brief flashes of bioluminescent organisms generating their own biological light. This is where we are taking the subs. Biologist Loren Eiseley was remarkably prescient when he wrote: "If there is magic on this planet, it is contained in water."

OPPOSITE: *Lights from* Mir Two *illuminate some of the spires and pinnacles at Lost City. Giant columns of calcite, created over thousands of years, rise from the summit of the Atlantis massif, and their vent fluids support lavish layers of microbes. Like other vent sites visited during the expedition, Lost City may hold clues to the possibility of life on other worlds.*

Marine biologist Dijanna Figueroa (ABOVE) marvels at the blue that surrounds her as *Deep Rover 1* begins a one-hour descent to the bottom of the Atlantic. Working at night in high winds, the Russian sub crew scrambles to recover *Mir One* from the ocean (LEFT).

Basket stars, a type of
starfish closely related
to brittle stars, cluster
on a rock at Snake Pit
in the Atlantic Ocean.

Working in tandem at
nearly 3,000 feet,
*Mir Two* illuminates a
Lost City spire as *Deep
Rover 2* moves in for
a closer look.

# CHAPTER TWO

# ON THE HIGH SEAS

## EXPLORING AND FILMING

Minerals such as iron, copper, and zinc sulfides in superheated hydrothermal vent fluid react with cold water, creating the "smoke" blasting from this black smoker vent.

## AUGUST 2003

*"Photo Yuri," a Russian crew member, snaps a predive portrait of the crew of* Mir One.

Early in the morning of August 1, after steaming from Newfoundland and Fort Lauderdale via the Bahamas, the *Keldysh* and the *Ares* rendezvoused in the Atlantic Ocean at 37° north latitude. A canopy of light gray clouds occluded the sun. The temperature was in the mid-70s and a gentle swell was running from the southwest. Some of the islands of the Azores lay 120 miles to the north.

The crews of the two ships exchanged visits, inspected each other's subs, discussed the dives they were about to make, and then went to work. The *Mirs* had already made five dives to the Menez Gwen vent field, 3,000 feet below the keel of the *Keldysh*. Cameron's plan was to have the four subs dive together at Menez Gwen, and then steam 450 miles south to the vent field at Lost City and repeat the procedure. Then the two ships would move farther south so the *Mirs* could take the scientists and filmmakers down 8,000 feet to the site called Snake Pit. If the weather held—and hurricanes were known to arrive in these waters this time of year—the 20 dives could be completed in two to three weeks.

While waiting for the *Ares* to arrive, Cameron conducted a series of preliminary dives. The first dive was made in *Mir 1* to confirm the location and test the 3-D-HD camera mounted on the sub. When this proved successful, another dive was made on the next day. Jim Cameron, marine biologist Dijanna Figueroa, and Anatoly Sagalevitch dove in *Mir One*. Mike Cameron, Kevin Hand, and the pilot Victor Nischeta dove in *Mir Two*. Because it was a shallow vent system, the divers observed midwater animals including fish and crabs everywhere. Dijanna Figueroa later told me: "We dropped down through shimmering water and landed right on the spot!" Mike Cameron steered his robot over to a chimney and recovered a small sample of sulfide material. Sagalevitch used *Mir One*'s mechanical arm

OPPOSITE: *As the sun slides into the Atlantic, the crew of the* Akademik Keldysh *prepares for a night dive.*

to recover rock samples and mussels, and a corrosion sampler was deployed.

On the day the *Ares* arrived at Menez Gwen, *Mir One* and *Mir Two* made another tandem dive. A large squid swam very close to *Mir Two*, and both subs were surrounded by numerous animals, including boarfish, siphonophores and a "bearded necklace" creature.

On the first afternoon, with the sun burning like a nickel in a white-hot sky, a problem arose that threatened to sink the mission. While launching *Deep Rover* 2 over the stern to make its first dive in the Atlantic, the A-frame suddenly became catatonic. Jim Cameron and his brother J.D. walked across the searing heat of the deck, stood under the giant, cream-colored A-frame, and stared at a gleaming steel piston. The piston, a major component of the A-frame, was not moving. A few inches below its lower end was a ring of three-inch bolts. Something had happened inside the piston's hydraulic system that caused the bolts to strip their threads and bring the slow, overhead arc of the A-frame to a noiseless halt.

Both men were silent, their minds forming questions that would determine the outcome of the expedition. How serious was the problem? How soon could it be fixed? *What's the next step if it can't be fixed?* They talked quietly between themselves and then slowly drifted down the deck toward a knot of men wearing orange coveralls and hardhats.

The look on everyone's face said that getting subs to work and getting them into the ocean wasn't very cool. It was dirty, gritty, and sometimes heartbreaking. An hour later, the Cameron brothers and the deck crew had answers to the first two questions and were working on the third. The piston was frozen. They couldn't launch the *Deep Rover*s. Repairs couldn't be made anytime soon, meaning the A-frame was inoperable for the rest of the expedition.

Jim Cameron and the men on the *Ares* had been working nonstop on the A-frame launch system since they laid down the narrow-gauge track in Fort Lauderdale. They had figured out how to roll a sub along the deck and up an incline until it sat directly under the lift point. They had worked out a way to transfer the weight of the six-ton sub from its trolley to the A-frame. They had selected eight well-muscled men from the ship's crew and the sub crew and trained them to run two deck winches and pull on five lift-lines and hoist points. The entire procedure was written down and fine-tuned. With the *Ares* tied securely to the concrete pier at Port Everglades, they practiced and then practiced again.

After eight days in Fort Lauderdale, they steamed across the Gulf Stream to the Bahamas, moored the ship to the commercial dock in Nassau, and rehearsed some more. Finally, when they felt they had a procedure that worked, they went to a deepwater site off the west end of New Providence Island and began to dive the *Deep Rover*s in earnest. There, in the cobalt blue waters near a submerged escarpment, they discovered how flawed the subs really were. One after the other, the cooling systems quit, the communication systems failed, and the thrusters malfunctioned.

Undersea expeditions rapidly shape the men and women who participate in them. They face an obstinate ocean and work that is never finished. The toil makes them sweat in streams that slick their faces and plaster their T-shirts to their backs.

As the days blurred into nights, the sub crew mustered all the technical savvy they possessed in order to troubleshoot the *Rover*s. On their second attempt, both subs made the 3,000-foot dive that gave them the all-important American Bureau of Shipping certification. Hours before he caught a plane to St. John's, Cameron shot his first roll of 3-D-HD images from inside *Deep Rover 1*.

As a tropical storm dumped two inches of rain on Nassau, the sub crew left the ship, hoisted a few rums at a local bar, climbed back on board, and began the eight-day voyage across the Atlantic to Menez Gwen. While the *Ares* steamed along the great circle route that took them south of Bermuda, they spent every waking hour working on the subs. They replaced burned-out cables, rewired the thrusters, and repaired the air conditioners that cooled the interiors. Some nights they fell into bed with their clothes on, closed their eyes for an hour or two, and then shuffled back out on deck just as the sun was breaking over the ocean.

Far to the north, the *Keldysh* had left Newfoundland and was plowing south-southwest at a speed of ten knots. In the four days it took to travel the thousand miles to the Azores, the big ship's cabins and decks were filled with anxious men and women hurrying about on unfinished business. On the main deck, under two white retractable roofs, the Russian sub crew was swarming over *Mir One* and *Mir Two* with torque wrenches and replacement parts. Jim Cameron was thinking hard about the best way to mold two ships, four subs, four 3-D-HD cameras and 170 people into a cohesive team. Mike Cameron was cannibalizing critical parts from one of his robots to make the other one work.

In a well-lit room deep within the *Keldysh*, Ed Marsh was reviewing the footage shot in Fort Lauderdale and Nassau. Marsh is a core member of Jim Cameron's documentary team. In 1987, he edited sequences for *The Abyss*. Shortly after, he wrote, directed, and edited *Under Pressure: Making* The Abyss. Since that time Marsh has worked on almost all of Cameron's undersea productions in one capacity or another. On both the *Keldysh* and *Ares*, Marsh would sit in front of a bank of video screens and study the 3-D images streaming from camera cable feeds into nearby recorders, commenting on the material and making suggestions to the film team. Marsh also directed the in-depth interviews of the scientists that make up most of the film's voice-over narrative.

Scattered throughout the *Keldysh*'s seven decks were the scientists and science educators Cameron had invited to participate in the expedition. They included Michael Atkins, a microbiologist; Maya Tolstoy, a marine geophysicist; Megan McArthur, a NASA astronaut; Loretta Hidalgo, a biologist; and Christy Reed, a science journalist.

Atkins, a muscular 41-year-old with an easy smile, has a Ph.D. in biological oceanography from MIT and the Woods Hole Oceanographic Institution. As a graduate student at Woods Hole, he undertook the first-ever survey of protozoa living near deep-sea hydrothermal vents. Yet, he told me, all of his work had been done in the library and laboratory, and he was really excited about making his first dive in a research sub.

I interviewed Christy Reed as she stood in front of *Mir One*. Slim, dark-eyed, and 26 years old, she kept glancing over her shoulder at the array of science equipment mounted on its bow. When she was 18 she fell in love with the earth sciences, and for the past five years has been writing about marine geology, plate tectonics, and hydrothermal vent systems in journals such as *Scientific American*. Like Atkins, she couldn't wait to climb in a sub for her first close-up view of the world's least understood ecosystem.

All of the scientists on the big Russian ship had moments when they wondered what they'd gotten into. They were puzzled by the sounds and silences of

*First assistant director Ellie Smith and Creative Producer Ed Marsh discuss the film's progress in Mission Control, where all of the film's imagery comes for processing and editing. The glasses Marsh is wearing enable him to see the footage in 3-D.*

a language they could not understand. Some had never been to sea before and had mild motion sickness. Only one had made a dive inside the eyeball-tight combat space of a research sub. As they moved through the prison-lit corridors of the *Keldysh,* they looked like anxious characters in a B movie.

Each of them shared an unspoken loyalty to Cameron, his team, and their cause. They didn't know much about the acclaimed Hollywood movie director, but he clearly loved exploration and discovery and telling the stories to a wide audience. Whatever happened, they were delighted to be along for the ride.

*Sparks fly off the* Ares *as welders cut through the hull to make an emergency opening to side-launch the critical* Deep Rover *subs.*
*"When the Russians saw the sparks coming off the Ares, they decided we were all crazy," says Jim Cameron.*

After four hours the crew of the *Ares* confirmed the A-frame was beyond repair. At ten o'clock that night, in the glare of the lights next to the *Deep Rovers,* Jim Cameron announced his backup plan. His idea was so outrageous it made everyone smile.

"We're going to cut a hole in the side of the ship," he said. "We'll use the mid-deck ten-ton boom crane,

rerig the control lines, and launch the subs over the starboard side."

No one cuts a hole in the side of a ship, even if it is above the waterline, but Cameron had the weight of a 14-million-dollar expedition and film on his shoulders and there was no other option. The captain of the *Ares* was speechless, but it was his A-frame that had failed at the critical moment, so he nodded assent.

"Let's get started right now," said Cameron. "And let's film it. If I'm going to go through all this misery, we're going to get a scene out of it."

Two cutting torches were hauled out and hooked up to high-pressure cylinders. Two men in thick canvas coveralls and welder's faceplates ignited their torches and waited a few seconds until the tip of each torch reached 5,000 degrees Fahrenheit. Inside a shower of sparks, they began cutting into the one-inch steel plate. High above them, behind the slanting glass of the ship's bridge, the captain of the *Ares* covered his eyes and turned away.

By dawn the next morning you could stand on the deck and see daylight through a rough-cut opening the size of a school bus—big enough for a *Rover* to slide through with plenty of room.

During the next two days, Cameron spent hours in the Survey Room with a pencil and sketch pad trying to figure out how to jury-rig the boom, winches, and handling lines. He attacked the problem with his trademark intensity, calculating the physical forces that had to be controlled when six tons were hoisted up and moved sideways across the deck. He knew that,

if caught in a series of swells, the *Ares* was going to roll like a rum-soaked sailor.

Down on a deck filled with roaring machinery, men in hard hats and grease-stained coveralls manned lines and operated winches in ways unlike anything they'd ever done before. They used hand signals and occasional shouts. To a man, they were both infuriated and uplifted by their leader's ruthless energy.

The launch and recovery team initially tested the new system by hoisting up a wire cage full of equipment. A swell rolled under the ship, and the cage got away from them and slammed into the far side of the ship. They quickly made adjustments. They tried lifting an empty *Rover,* moving it sideways for a few feet, and then putting it back on its trolley. It worked, so they tried it again and almost made it across the deck to the water. Something jammed, so they rerigged the network of lines leading from winches to pulleys to pad eyes.

Late one night, Cameron and the men in orange coveralls learned how makeshift their system really was. As they lifted *Deep Rover 2* out of the water and inched it back toward the trolley, the wind came up, the ship began to sway, and the sub started to swing.

They pulled on the lines, but with a sound like a gunshot something snapped under tension and the sub crashed into the side of its trolley. There were curses and groans. In desperation, they dropped the sub on the deck. The sound of metal screeching on metal was followed by a long silence.

The struggle took its toll. Everyone was exhausted. There were bruised shins, a twisted ankle, and patches of scraped skin oozing red in the night.

Cameron consulted the weather forecast and saw that a storm front was moving in from the west. He decided not to dive the *Rover*s at Menez Gwen. Instead, they would steam south to Lost City. Further work on the launch and recovery problem could be done while they were under way.

For the next 30 hours the *Ares* and *Keldysh* steered a course of 227º, keeping within sight of each other. High winds and heavy seas forced them to slow to less than ten knots. At dusk on the first night, the Greek captain of the *Ares* stood on the bridge, eyed the white spray cascading over the bow, and said, "This is why it's called groundswell. Feels like the whole damn Earth is shifting beneath your feet."

As the winds increased and the sky turned black, Cameron went up to the washed-out light of the Survey Room and hunched over his yellow notepads. He wasn't thinking about the film he was supposed to be making. There would be no film—at least the way he wanted to make it—without the *Rover*s, so he calculated lift angles and estimated righting moments that would make the launch and recovery more efficient. He'd been on deck for every practice launch, in the thick of things, hauling lines, letting the kinetics of the problem go into his brain, and he was confident he could find a solution that would work.

He has the answer by the time we reach Lost City. A brilliant morning sun is streaming through the window blinds of the Survey Room on the second deck of the *Ares*. Inside the L-shaped room 20 people are sitting around a long wooden table, leaning against the walls, or sprawled on the floor. Some are taking notes; most are looking at Jim Cameron, wearing a gray T-shirt with MARINES printed on the front and talking about the dive plan for the day. In the far corner of the room a man in his mid-30s is blinking hard, trying to keep his eyes open.

On the table are two yellow sketch pads, a black pencil, and four small-scale models of the two *Mirs* and the two *Deep Rovers*. "This is the first time four subs have ever worked this deep," says Cameron. "So this dive is going to test all our communication, navigation, and piloting skills."

There are dark circles under Cameron's eyes. He's been awake since five A.M, working on the choreography of today's four-sub dive. At 7:30 he held a meeting on the *Keldysh* with the scientists, the 'bot team, and the *Mir* pilots. Then he climbed into a Zodiac and raced across to the *Ares*. The people in the room include his brother J.D., three scientists, the seven-man *Rover* team, and the *Ares* launch-and-recovery crew.

"The *Mirs* will be launched first," says Cameron. "They'll descend to the top of the massif and find the big carbonate towers we shot in 3-D last year. Then they'll try to find a rendezvous point near the towers. When they locate the area, they'll call the surface and tell them they've found a suitable spot. They'll give their best estimate of the direction and

*Minutes after hearing that the A-frame couldn't be repaired, Cameron began focusing on an alternative solution.*

speed of the midwater currents, so the *Rovers* can drop directly to the site."

Cameron picks up the two *Deep Rover* models and holds them side by side above the table. "Tym Catterson will pilot *Rover 2*," he says, "with Loretta Hidalgo as science observer. I'll be shooting in *Rover 1* with Patrick Lahey as pilot. It's critical we stay as close as possible during the three-thousand-foot descent."

As Cameron is talking, two of the men sprawled on the floor nod their heads, close their eyes, and fall asleep. With their sweat-stained T-shirts and grease-stained blue jeans, they look like survivors of a shipwreck who have lived through the most demanding days of their lives.

Cameron brings the *Deep Rover* models close together and places them side by side on the table. "We'll launch *Rover 2* at 1200 hours," he says. "It will remain on the surface until *Rover 1* is in the water. Once we've confirmed we're ready to dive, we'll leave the surface together."

Cameron searches the room with his eyes until

# THE MID-ATLANTIC RIDGE

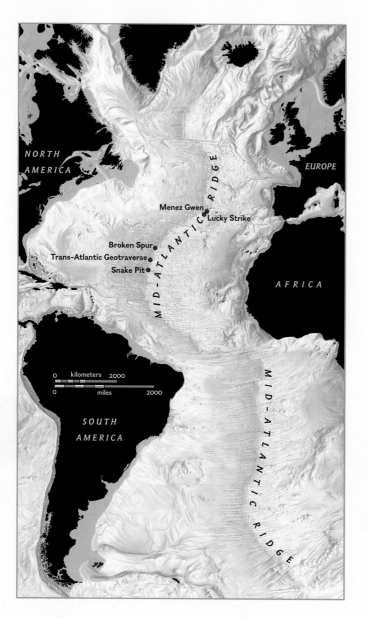

The Mid-Atlantic Ridge is a submerged chain of volcanic mountains with a central rift valley that bisects the entire length of the Atlantic Ocean. At the bottom of the rift valley molten rock, or magma, rises from deep within the Earth's interior. The magma cools, solidifies, and spreads, creating new seafloor that widens the Atlantic at a rate of about one inch per year. This mountain- and ocean-building process has been under way for nearly 200 million years.

Because it is constantly spreading, the seafloor at the bottom of the valley is filled with cracks and fissures that allow cold, dense seawater to seep down and encounter the superheated magma beneath the crust. In this highly pressurized environment, sulfate, which is abundant in seawater, is converted into sulfide, and additional minerals leak from the magma into the water. Expanded by the intense heat, the water pushes back up toward the rift valley. Where it breaks through the crust and into the ocean, a hydrothermal vent is formed.

Depending on the variations of rock structure and heat, hydrothermal fluids emerge as warm shimmering streams, smokelike white plumes, or belching black jets with temperatures in excess of 700°F. High temperatures, high acidity, and high levels of hydrogen sulfide characterize most of the vent fluids studied so far. Each of these chemical characteristics has far-reaching biological implications. One of them is that the very limited zone where vent fluids mix with oxygenated seawater can become densely populated with creatures thriving directly or indirectly on a diet of hydrogen sulfide.

Some of the most dramatic vent systems are the superhot black smokers. The collision of their mineral-rich water with the near-freezing ocean water forces the materials in the plume to form black "smoke." The particles that make up the smoke are the polymetallic sulfides of minerals like zinc and copper. When they cool and solidify, they create the contorted black chimneys beneath the smokers.

The rate of seafloor spreading and the behavior of the underlying magma chamber determines whether there are biological communities at a vent site. Where spreading is slow, chambers can cool and the animals can die off. A hydrothermal vent teeming with a biomass 500 to 1,000 times denser than the biomass of the normal deep sea can be radically altered by a single volcanic eruption.

Along the Mid-Atlantic Ridge, scientists have studied a number of vent fields, including:

- **TAG** (Trans-Atlantic Geotraverse). Black smokers, white smokers, and swarms of blind gray shrimp are found at this site, located at 12,000 feet.
- **Menez Gwen**. Sitting on the summit of a volcano, this is one of the shallowest sites, just 3,000 feet below the surface. Mussels, crabs, and at least 33 other species have been observed here.
- **Lucky Strike**. The largest hydrothermal area known, Lucky Strike covers 90 square miles and has 21 active chimney sites at a depth of 5,600 feet.
- **Broken Spur**. With a similar biology to TAG and Snake Pit, Broken Spur, located at a depth of 10,000 feet, is dominated by dense swarms of shrimp.
- **Snake Pit**. One of the most biologically rich vent sites in the Atlantic, Snake Pit is located in 12,000 feet of water and is home to dense fields of mussels and swarms of shrimp.

OPPOSITE: *The Mid-Atlantic Ridge is split by a rift valley three to five miles wide. In places its walls are higher than the Grand Canyon.*

finds the two *Deep Rover* pilots. "If one *Rover* experiences a horizontal thruster failure during descent, the other will maneuver to within 100 feet of it—until we reach the bottom. If the bottom depth is beyond 3,000 feet, the disabled sub will abort the dive and the other sub will continue. The other sub will be given vectors from surface control and proceed to rendezvous with the *Mir*s."

He reaches down and picks up a *Mir* model with his left hand and a *Rover* model with his right.

"*Mir One* and *Rover 1* are the 3-D-HD camera platforms. *Rover 1* is the primary camera platform, and its shot will take precedence. If *Mir One* gets a good alignment of subject and lighting, Vince Pace will say he has the shot and *Rover 1* will go into the supporting role. *Rover 2* will be the primary on-camera vehicle. *Mir Two* will primarily provide lighting support. Its secondary role will be to take science samples."

Cameron pauses and takes a deep breath. "In the event of a loss of communication between the *Rover*s

*Jim Cameron explains the complicated four-sub dive at Lost City as his brother Mike looks on.*

and the surface, or a loss of communication with the *Mir*s, the *Rover*s will commence a one-hour waiting period. All other systems on the sub must be operational with no threats to safety. If there is no communication after one hour, the *Rover*s will abort the dive. If either *Rover* loses communication, it will terminate the dive. The other *Rover* may remain and continue to work."

These are crucial words, and everyone in the room is leaning forward. "When the work is completed or if the battery power is low—or if technical performance is compromised in any way—the *Rover*s will begin to ascend. *Rover 2* will ascend first. When it has been recovered on board, *Rover 1* will be cleared to ascend. Neither sub will come up without clearance from surface control."

The men and women have arrived at one of the moments they have been working toward for five long months. After Cameron's last words, the room empties in less than a minute. All that is left are the sketch pads on the table and the sunlight slashing through the slats of the blinds.

A few hours later, Cameron climbs into *Deep Rover 1* and takes a seat next to the pilot. His eyes are narrowed with concentration and his forehead gleams with sweat. He says a few words into the intercom, checks the monitor of the 3-D-HD camera and glances across the deck. Finally, on this technically spooked expedition, everything seems to be working. Even the ocean is flat as oil on a griddle.

The pilot, Patrick Lahey, finishes his predive checklist and signals to the deck crew they are ready for launch. Standing in the middle of the deck, J.D. Cameron motions to the crane operator. With a gentle upward tug, the sub is lifted off its trolley and moved sideways. Eight men tending control lines guide it across the deck. The air is filled with diesel exhaust, the sounds of winches and capstan engines, and an occasional shouted order. Every few feet the sub is brought to a stop to ensure the control lines are under the right tension. As the sub and its two occupants are lifted over the side of the ship, a towline is tossed from an inflatable boat and attached to a pad eye on the sub's port battery pod. The inflatable boat turns and pulls the towline taut to keep the sub away from the ship. On J.D. Cameron's

command, *Deep Rover 1* is dropped smoothly into the ocean.

J.D. walks to the railing and looks down at the half-submerged sub. "That's the easy part," he says to the man standing beside him. "The hard part will be getting it back on board."

As *Rover 1* bobs on the surface, Jim Cameron and Lahey are clearly visible through the transparent pressure sphere. They are wearing blue coveralls and headsets. Both are checking the display panels and instruments in front of them.

Cameron has made more than 40 dives in the 18-ton *Mirs*. "They're deep-sea workhorses that'll take you to 20,000 feet for eight hours," he likes to say. "But their small ports severely restrict what you can see." As soon as *Rover 1* is completely below the surface, the veteran of dozens of dives to the *Titanic* and *Bismarck* sweeps his eyes full circle around the crew-sphere. It's a moment he's been anticipating for months—his first panoramic view of the inside of the North Atlantic Ocean.

A diver jumps out of the inflatable boat, swims over to the sub, and releases the crane's lift hook. Inside, Lahey and Cameron look up and see *Deep Rover 2*, launched less than an hour ago, moving slowly toward

them. Lahey talks to pilot Tym Catterson in *Rover 2* and then says a few words to the surface control team. He glances upward at the sun fixed in the sky.

"We're beginning our descent," he says quietly. And with those words, Jim Cameron begins another undersea shoot on the world's biggest movie set.

Cameron and the scientists have made three tandem dives in *Mir One* and *Mir Two* at Menez Gwen. Less than 3,000 feet from the surface, the site was surrounded by a swarm of midwater creatures including siphonophores, boarfish, and a large squid. They filmed the vents and collected biological and geological samples including a vent mussel called *Bathymodiolus azoricus*. Tissue from some of the mussels was frozen so its heat-shock proteins could be analyzed on board the *Keldysh* and at a university laboratory in California.

But this dive is different. On this fine, sunny afternoon, with the ocean eerily calm, Cameron is attempting something no one has ever done before—directing four subs down to a depth of 2,800 feet, navigating through the blackness toward a mountain almost as high as Mount Rainier, locating its summit, and then landing on a small, sloped terrace in a vent field called Lost City.

The hydrothermal field, named for its breathtaking white limestone towers, is 30 degrees north of the Equator. The field and the ocean-floor mountain beneath it called the Atlantis Massif were discovered in 2000 by a team led by geologist Debbie Kelly.

Lost City rests on a terrace on a south-trending spur of the massif. It extends more than a thousand feet across the terrace and has at least 30 active and inactive vent structures. It hosts spires, mounds, and pinnacles, many of which are 30 to 90 feet tall. The most spectacular is a giant circular tower the height of an 18-story building. "Poseidon" is the tallest hydrothermal deposit yet discovered. Four small spires venting fluids mark its tortured limestone peak.

As they fall through the sunless layers of blue-green and black water, the lights on both *Deep Rover*s are switched on and Cameron the director becomes Cameron the cinematographer, controlling his camera by turning small hand wheels that adjust its pan and tilt. He commands Lahey to fall in behind *Rover 2*, calling out the depths and positions he wants with the impatient voice of a man

OPPOSITE: *Tym Catterson and Kevin Hand carefully monitor* Deep Rover 2's *descent to a vent site.*

who sees every scene long before he shoots it. Yesterday, *Rover 2* aborted its dive because a thruster and its communication system failed. Today, Cameron is hoping that the subs will keep running long enough to get some big-scene footage. Tucked away in the back of his mind is the fact that if the titanium housing holding the 3-D camera—positioned on the outside of the sphere six feet away from his face—were to implode under pressure, the shock wave would shatter the sub's sphere and in less than a second he and Lahey would end up looking like strawberry jam.

As the subs continue falling toward the center of the Earth, *Rover 1*'s altimeter, which monitors the sub's depth, stops working. A few minutes later, what Lahey describes as "a spectacular power outage" occurs. The sub goes black, becomes "a dead boat," and automatically switches over to its emergency batteries. Lahey, a skilled pilot, immediately finds the right panel, flips a few switches and brings the communication, hydraulic, and thruster systems back to life. Through all of this, the wide-angle "witness" camera inside the sub shows Cameron smiling and aiming his 3-D camera at the other *Rover*. Like Lahey, he has lived through many adrenaline-charged moments, and this isn't one of them.

Forty minutes after leaving the surface, they are at 2,500 feet, using the other sub's altimeter to guide them to the summit of the Atlantis massif, a 14,000-foot dome-shaped structure that is nine miles away from the center of the Mid-Atlantic Ridge. The first thing they see is a steep slope of gray-white rocks. They slow their descent until they are hovering just above a line of boulders. All around them, hidden in the silky shadows, are the spires, mounds, and pinnacles of the hydrothermal vent field. Some pinnacles are extinct; others are active. The tops of the thermally active pinnacles shimmer with a mixture of hot vent fluid and cold seawater.

Three hours before, the two *Mir*s were launched from the nearby *Keldysh*. *Mir One* carries pilot Anatoly Sagalevitch, cameraman Vince Pace, and science journalist Christy Reed. *Mir Two* carries pilot Genya Chernyaev, engineer Mike Cameron, and marine geologist Maya Tolstoy. Tucked inside its enclosure on the bow of *Mir Two* is Mike Cameron's robot. The *Mir*s have found a site and are somewhere below them, waiting.

As Cameron and Lahey watch, a large deepwater fish swims out of the gloom into the glare of their lights. It is a wreck fish, more than a hundred pounds of eyes, scales, and tail. It comes to a stop in front of

*Mike Cameron's marvel of deep-sea engineering,* Jake *the robot, makes close-up photographs of vents and animals that are beyond the reach of the expedition's four research subs.*

the sphere, stares at them with unblinking eyes, and slowly turns away with an air of disdain. "Checking us out for a meal," laughs Cameron.

As soon as they get their bearings, they start moving across the slope above the pinnacles and then between them, searching for the lights of the two Russian subs. Even with beacons on all four subs and a constant stream of directions from the surface control team on the *Ares,* finding the *Mir*s is a difficult task. The three-mile-high massif has a tangle of scarps and spurs on its south face that are swept by unpredictable currents. At one point they fly both *Deep Rover*s, one behind the other, through a gleaming white opening in the rocks that looks like an inverted Arc de Triomphe.

# LIVING AT EXTREMES

For the human body, the conditions at a black smoker vent field two miles under the ocean are lethal. The darkness lasts forever. There is no oxygen to breathe. The pressure is more than 5,000 pounds per square inch. The water next to the vents is as cold as death, while the fluid jetting out of the mouth of the chimneys is hot enough to melt lead. And the water is filled with sulfuric plumes of polymetallic minerals. Yet diverse communities of animals live—indeed thrive—in these conditions.

Scientists have identified more than 500 unique species of animals at hydrothermal vent sites around the world. Many of these were completely unknown before the discovery of the vents in the late 1970s. The shrimp, mussels, crabs, and tube worms that dwell on the vents have not only adapted to their harsh environment, they embrace it.

Within a few inches of a chimney's 650° F orifice, the temperature is a comfortable 75°F. A few feet away it is 35° F. In a place dominated by severe thermal and chemical gradients, the animals have adapted to move or secure themselves in nearby niches where the temperature and chemistry enhance their physiology. But most of their reflexes and instincts are designed around the continuous search for something to eat.

The availability of food is the primary limiting resource for all deep-sea organisms, especially those dwelling on the vast, soft-bottom habitat that covers most of the world's oceans. Only a small percentage of the food energy produced by photosynthesis in the surface waters is transferred down to the deep-sea floor, where conditions are uniformly cold and relatively stable. The result is that the population densities and biomass of most deep-sea animals, including brittle stars, grenadier fish, and sea snails, are low. Like their counterparts in shallow cold-water ecosystems, their metabolism, growth rates, and colonization are relatively slow compared with those of warm-water animals.

In marked contrast to immense deep-ocean areas such at the Atlantic Basin, hydrothermal vent fields are extremely small, discontinuous, and unstable. Biologically, they are high-density "oases" with elevated metabolism, concentrated biomass, rapid growth rates, and a limited number of species.

Most of the primary food energy at these vent communities is produced by more than 250 different strains of bacteria. In addition to using hydrogen sulfide discharging from the vents, some bacteria are capable of oxidizing iron, ammonia, and manganese. A few are capable of metabolizing in oxygen-free hydrothermal water. Scientists believe that these recent discoveries about the diversity of chemosynthesis greatly enlarge the range of conditions under which life could have originated on Earth—and may exist in other parts of the solar system.

There are a number of reasons why life has adapted to the vent fields. One of the most important is that some bacteria have symbiotic relationships with dominant invertebrates, including the giant tube worm and the giant clam. In a symbiotic relationship, two dissimilar organisms live intimately together in a mutually beneficial way.

The giant tube worm, *Riftia pachyptila,* found in the Pacific Ocean, grows to a length of eight feet inside a

OPPOSITE: *More than 500 species, including this vent crab, have been identified living in vent-system communities around the world. Many were unknown to science before the discovery of the hydrothermal vents.*

six-foot-long tube. The tube is as thick as the handle of a baseball bat and is composed of a tough substance called chitin, the same substance found in fingernails. The life of a tube worm is made possible because it feeds bacteria and bacteria feed it.

This is how it works. The tube worm has no eyes, no mouth, and no digestive tract. However, within its long body, inside a special tissue called the trophosome, is a swarm of bacteria—ten billion for every gram of tube worm tissue.

The tube worm is anchored to the bottom a discrete distance away from the superheated water where moderately warm currents mix with ocean water. Projecting out of its tube, the worm's delicate, feathery plumes gather in oxygen, carbon, and hydrogen sulfide. The tube worm's blood, rich in hemoglobin, has an affinity for oxygen and sulfide and carries these metabolites to the trophosome and the bacteria. The swarm of spherical microbes converts these raw materials into the carbon compounds that nourish the cells of the tube worm.

The large white clam, *Calyptogena magnifica,* also found in the Pacific, grows to a length of more than ten inches. Its rough foot extends deep within the cracks of the basaltic substrate, where the temperatures are as high as 600°F and the water is full of bacteria. The clam's soft tissue contains intracellular hemoglobin that transports oxygen to the symbiotic bacteria in its gills.

The two-inch-long *Rimicaris exoculata,* found on Atlantic black smokers, does not host symbiotic bacteria. Instead, these shrimp eat them by scraping the sulfide surface of the chimneys with the file-like spines on the tips of their legs and ingesting the bacteria-laden minerals. Their metabolism allows them to digest the bacteria and eliminate the undigested minerals.

These shrimp have developed a unique way to "see" in their pitch-black world. Two reflective lobes on their backs contain a chemical similar to the light-detecting pigment in human eyes. The shrimp can't see, but biologist Cindy Lee Van Dover believes they are able to sense the infrared wavelengths given off by the searing water. Although undetectable by human eyes, this thermal glow may guide the shrimp to the food-rich chimneys. Because 600° F water has a brighter gradient than 90°F water, they may use these gradients to avoid being turned into black smoker bouillabaisse.

OPPOSITE: *Multiple frames of film were interlaced to create this startling mosaic of a spewing black smoker, shot at the Snake Pit site in the Atlantic Ocean.*

"This is fantastic," says Cameron. "This kind of flying is impossible in a sub with small windows."

For years Cameron has been dreaming of using a pair of transparent subs—one as a light-sub, the other as a camera-sub—to film the interior of the ocean. Like a helicopter, the *Rover* allows him to see where he wants to go, turn on a dime, zoom in, and let the audience share the reactions of the explorers in both subs.

At 2,700 feet, the power fails again, and the lights of *Rover 1* dim and go out. In the blackness, Lahey looks down and catches a glimpse of the two *Mir*s. He points out the faint glow to Cameron and then reactivates the lights. Cameron tells *Rover 2* to go ahead and continue to descend. The two small subs, one above the other, drop down a boulder-strewn incline next to towering pinnacles of cream-colored rock.

As they descend, they see the two *Mir*s, their bows almost facing each other on a curving ledge. The stern of *Mir One* hangs out over the abyss. Their powerful lights, mounted on racks above their view ports, illuminate the area in front of them. Below their pipe-frame skids, a slope of shattered rock trails off into the darkness.

Cameron directs Lahey to hover so he can film *Rover 2* dropping in front of the two *Mir*s. When its gleaming sphere comes to rest on the broken rubble, all three subs are facing each other inside an intense halo of light.

"This is the money shot," Cameron keeps saying. He reaches down under his seat to check that the 3-D-HD decks are working. "This is when you make sure you're recording," he tells Lahey. "Otherwise, you'll want to shoot yourself."

He knows how fortunate they are to get these images. A platoon of people on two ships exhausted themselves for months to make the subs and cameras work. They needed a calm ocean to launch the *Rovers* smoothly. In spite of the currents and darkness, the American and Russian pilots have found their way to the same spot on the massif.

As Cameron focuses his lenses on the subs in the primordial wilderness below him, Mike Cameron's dark blue robot eases out of its enclosure on the bow of *Mir Two,* takes a short look around, and makes its way downslope toward *Rover 2.* Spooling out its thin white tether, it sees the sub, stops, and hovers in front of the crew-sphere. Its bow moves up and down as if it were waving hello. Inside, Loretta Hidalgo waves back.

The four subs spend more than two hours on the bottom. The *Mir*s and *Rover 2* remain in position while *Rover 1* hovers above them. Cameron calls a stream of instructions into the intercom. The robot takes close-up shots of the two people inside *Rover 2* and then flies down the slope below the subs and captures stunning close-up images of a white, fan-shaped creature leaning out from a cliff. Like Menez Gwen, Lost City is a shallow vent site and has many mid-water animals. On the coral plateau the team observes a number of interesting creatures, including a sea urchin moving on its tube feet, and white sea whips and gorgonians. Pairs of huge "grouperlike" fish are seen. Dijanna Figueroa would later comment: "Lost City was spectacular because there were sheer walls everywhere and funky creatures like a white crab with up-thrust legs, and fish living in holes in the bottom with their eyeballs sticking out."

Inside each sub, a small, wide-angle camera is recording every word and gesture. Vince Pace is bent over the controls of the 3-D-HD camera in *Mir One*. Mike Cameron is trying to coax his robot around an outcrop and back toward *Mir Two*. The two Russian pilots are silent, their eyes looking down at the gauges and then back out the view ports. Occasionally there

*The ROV Jake leaves its enclosure on the bow of Mir Two and heads towards the other subs.*

is a short exclamation: "I can't believe those lights. The ocean looks like it's on fire." Everyone is grinning. Even old-timers are acting like first-time tourists at the Taj Mahal.

Loretta Hidalgo is sitting in *Rover 2* with flushed cheeks and a smile on her face. She has overcome the fear of making her first dive in a mini-sub. She's safely on the bottom, and a few feet away there is so much high-tech hardware it feels like a traffic jam on the Santa Monica Freeway. Framed in the view port of *Mir Two* is her new friend Maya Tolstoy.

Tolstoy is a blue-ribbon scientist who uses seismic instruments to study undersea earthquakes, but on this dive she is once again the young girl in high school who loved discovering things. She can't believe how much life teems in the water around the subs. The *Rover*s look like science fiction UFOs. The sheer-walled carbonate spires seem like they are from another planet. "They should have sent a poet," she muses.

The pale columns and chimneys behind the subs, as high as the stacks of an ocean liner, stand passively in the artificial sunlight. They are formed as alkaline seawater seeps through a porous type of subcrustal rock called peridotite, heat is created, and calcium carbonate crystalizes. The vent fluids issuing out of the spires are relatively cool (104°F to 167°F) and support lavish layers of microbial communities. However, other than these dense mats of bacteria and the occasional wreck fish and siphonophore, life in the "city" is sparse.

Debbie Kelly and her colleagues believe Lost City is unlike any other vent systems found so far. First, there is the dizzying height attained by some of its structures. Second, the new vents are nearly 100 percent carbonate, the same material as limestone in terrestrial caves, and range in color from a clean white to cream or gray.

But Lost City's most distinctive feature is that it sits on 1.5 million-year-old crust nine miles away from the Mid-Atlantic-Ridge—leading to the possibility that many more such systems exist in other parts of the ocean. Within a 50-mile radius of the four subs are three other sites that may have the same fracturing and intrusion of seawater. And they represent a tiny segment of potential sites along the Mid-Atlantic Ridge.

If this kind of vent is common throughout the oceans, the oceanic crust may support far more microbial life than previously thought, increasing the amount of the Earth's biomass by as much as a third, according to some estimates.

Loretta Hidalgo looks up at *Rover 1* and the people inside the other subs. Her face wears a look of delight and uncomprehending awe. "We are the trespassers here," she thinks, "down here we are the extreme life."

It is the first time that four subs—50 million dollars of undersea technology—have ever worked together in water this deep. But the historic accomplishment contains another milestone. It is the first time humans have had a panoramic view of a hydrothermal vent system. The four people in the *Rover*s can look through the pressure hulls of their subs—up, down, and sideways—and get a 320° view of the ancient ocean. For each one of them it's like they have flown through the darkness and landed on a different planet.

OPPOSITE: *A Deep Rover collects a sample from the "Poseidon" spire at Lost City, the only vent site known that is not powered by heat from magma chambers. Instead, a chemical reaction caused by the interaction of seawater with mantle rock powers the system. The reaction creates unusual carbonate structures.*

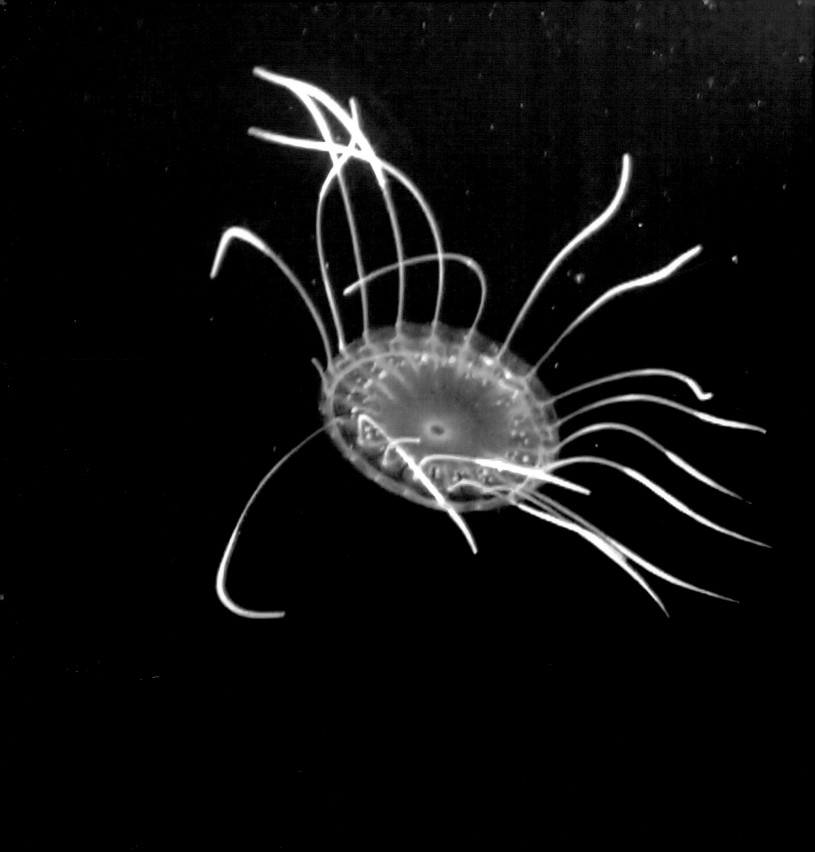

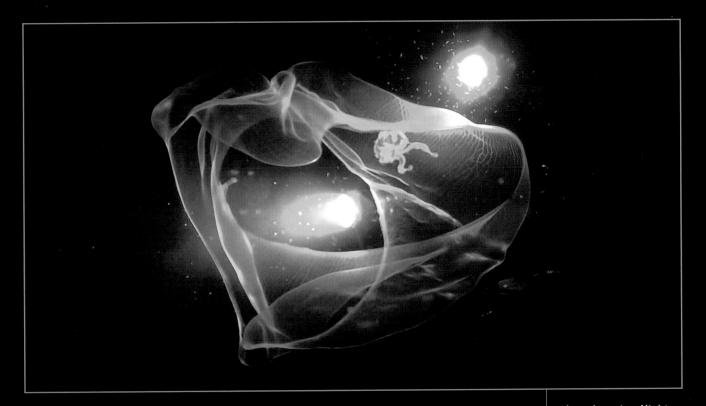

The submarines' light illuminates the spindly tentacles of a Solmissus jellyfish (LEFT) in an unearthly blue glow. The amorphously shaped jellyfish *Deepstaria enigmatica* (ABOVE) is often ob-served with an isopod crustacean—likely a parasite—within its bell.

*Mir Two*, with planetary scientist Kevin Hand on board, moves into position to collect a sediment sample from a small vent chimney at Menez Gwen.

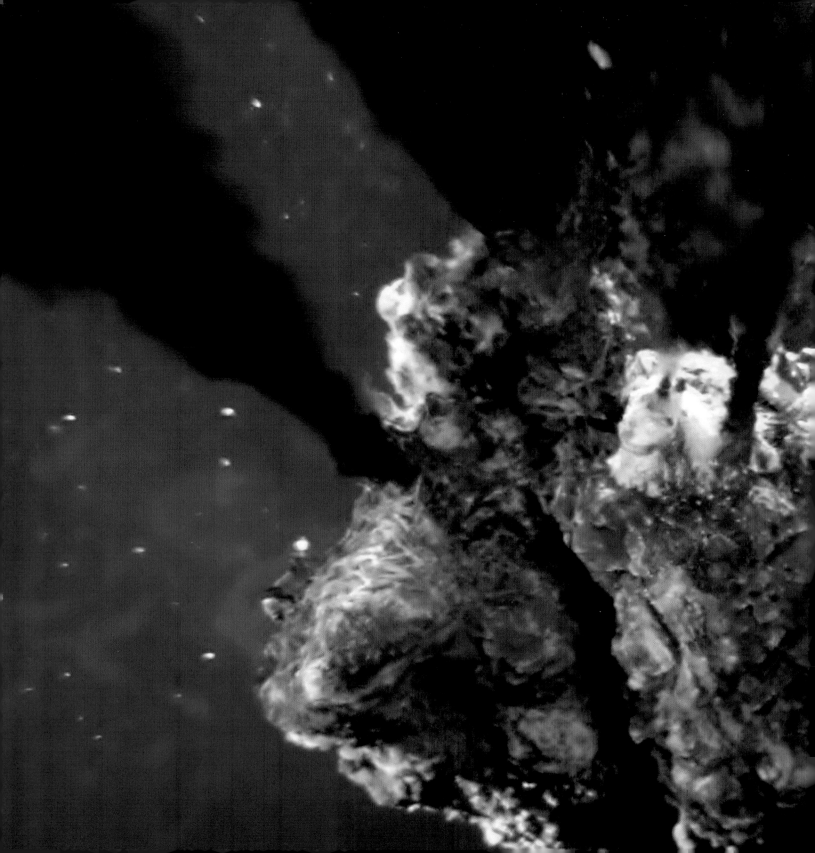

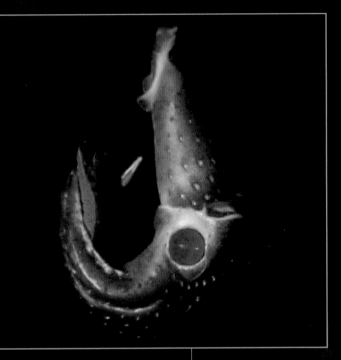

LEFT: Like an immense pipe organ played by the Earth itself, super-heated water shoots from a chimney. Such heavy venting activity indicates the presence of magma just below the surface. ABOVE: A very small squid floats above a chimney at Menez Gwen.

Loretta Hidalgo greets *Jake* with a wave as the 'bot glides in to examine the crew aboard *Deep Rover 2* at Lost City.

With a leg span of up to
31 inches, spider crabs
are a common predator
at hydrothermal vents.

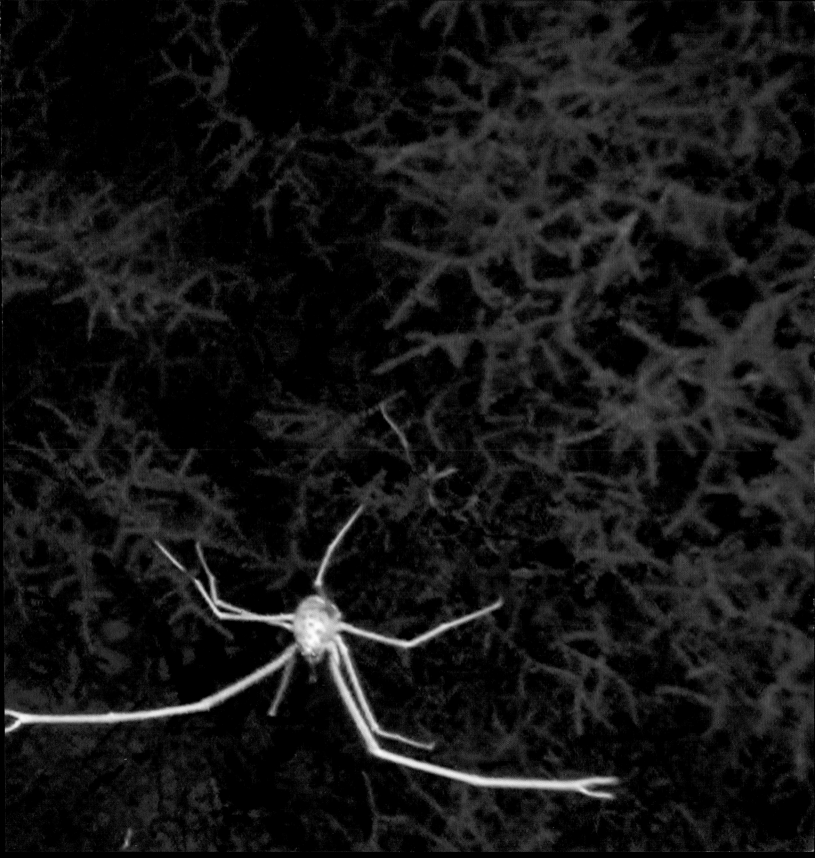

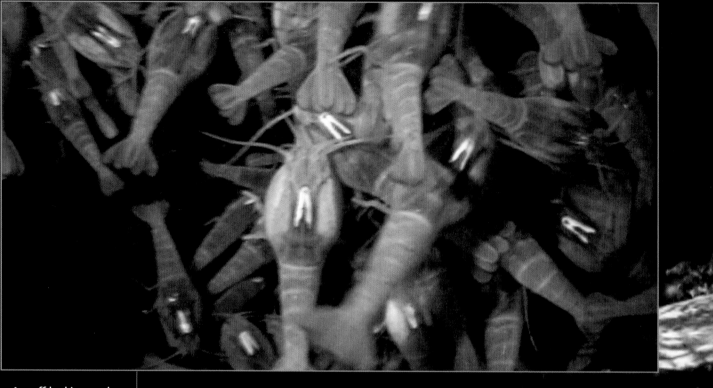

A gruff-looking angler-fish (RIGHT), "the ugliest fish in the world," according to Jim Cameron, is no match for the least tasty shrimp (ABOVE). Scientists who sampled *Rimicaris* report that the crustaceans taste like rotten eggs—a result of hydrogen sulfide in the vent fluid.

Testament to the great
abundance of life at the
vents, shoals of shrimp
cover the flanks of a
vent system known as
TAG, one of the largest
known sites in the
Atlantic. The shrimp
have evolved to thrive
at the vents, where tem-
peratures exceed 350°F.
The shrimp have no
external eyes. Instead,
scientists believe that a
pair of internal organs
helps the shrimp detect
radiant heat and very
low-level illumination.

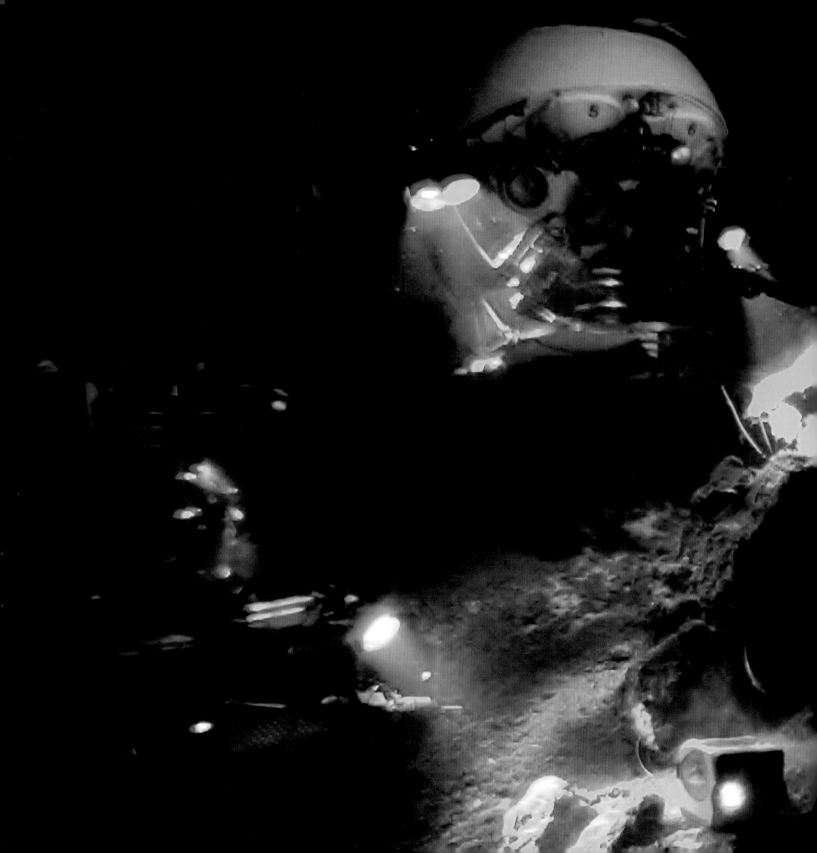

Like an expedition to deep space, four subs convene at a depth of nearly 3,000 feet at Lost City. It is the deepest four subs have ever worked together.

# CHAPTER THREE

# FROM SEA TO SPACE

## OCEANS OF LIFE

Io, one of Jupiter's more than 60 known moons, floats above the planet's spectacular cloudtops in this image from the Galileo probe. Several of the Jovian moons may hold clues for scientists searching for life in our solar system.

AUGUST 2003

During the long days and nights on the *Ares* and *Keldysh* small groups of men and women—sailors, sub pilots and technicians—congregate for conversations. They stand in the shade next to the *Mir*s and talk about the chores they need to complete before the next dive or problems they are having with a reluctant piece of equipment. They sit in workshops and laboratories and discuss the husbands and wives they left on shore, the weather coming over the horizon, their last film shoot, and their need for a good night's sleep. In spite of their unusual reserves of hardiness, the round-the-clock rhythm of diving and filming and maintaining their gear is exhausting.

The scientists among them talk about the samples that have just come to the surface on the *Mir*s, the journals they are reading, the challenge of raising grant money, and the unfinished work waiting for them in their universities and offices on shore. At times, especially late at night, their talks turn from the subject of Earth's ocean to its connection to life in other parts of

*The wide-eyed rattail, or macrourid, fish is an occasional visitor to the vents, while the clams visible in the background are permanent residents.*

the solar system. These lines of thought often turn into discussions about exploring Mars and the moons of Jupiter.

On one hot, sun-bright day a group of them are having lunch with Jim Cameron in the *Keldysh* dining room. The air is charged with the deep blue shimmer of the ocean just outside the portholes. Everyone's voice, including Cameron's, is infused with the excitement of recent dives and personal discoveries.

Cameron turns to Kevin Hand and asks him what the most interesting vent site was that he's seen so far. Hand, the geologist turned astrobiologist, had done his homework before joining the expedition. Like everyone around the table, he had scanned the websites, talked to colleagues, and read the available literature. But nothing prepared him for the reality of staring into the shimmering throat of so many different kinds of vents.

OPPOSITE: *Hot water streams from a black smoker in this early photograph of a tubular hydrothermal vent.*

He answers without hesitation. "Snake Pit was the most incredible because of the size and shape of its black smokers. I still can't get over the Dr. Seuss-like morphology of the spires, chimneys, and flanges."

Because she is a marine biologist, Dijanna Figueroa saw Snake Pit from a different angle. "I was knocked out by the biomass," she says. "The number of shrimp around the chimneys was mind-blowing."

Jim Cameron clears his throat. He's made dozens of extreme-depth dives in the Atlantic and Pacific, and he's spent more time on *Titanic* than the ship's captain, Edward Smith. "I'd have to agree with you. There was more going on there than in any place I've seen underwater."

As I listen to their conversation, I realize they talk about these subjects with such authority because they live at the precise moment in history when humans for the first time can dive deep into the ocean and sail off into the sea of space. In recent decades these remarkable parallel journeys were made in spacecraft and seacraft with names like Mercury, Vostok, Mariner, Apollo, *Alvin* and *Trieste*. They changed forever the way we think about the surface of the moon and the planets and altered our perceptions of the floor of the ocean and the life that thrives there.

The sea-space journey began in the early 1960s when Yuri Gagarin orbited the Earth in Vostok 1 and Lt. Don Walsh and Jacques Piccard dove seven miles into the Pacific Ocean's Mariana Trench in the bathyscaphe *Trieste*. In the 40 years since, several major milestones have been achieved.

One of them took place in the spring of 1979 when marine biologists in the research sub *Alvin* had their first close-up look at the Galápagos vent site discovered two years earlier. Unlike the geologists who found it, they came prepared to study, photograph, and sample the alien horde of primary producers, symbionts, grazers, and carnivores that lived in the chemically corrosive water. As the dives progressed, the ship's laboratory filled with preserved samples of creatures new to science.

After these dives, the biologists dove to a hydrothermal vent site near the Gulf of California called 21° North. At a depth of 8,000 feet, they looked through the view ports and saw clusters of mussels and tube worms and crabs within this deep-sea oasis.

OPPOSITE: *Expecting to find a cratered, dead world, NASA scientists were astonished to discover that Europa was relatively young, with a surface of shifting ice.*

And they glimpsed something no one had ever seen before—black smoke boiling out of a tall, corrugated stovepipe. They studied it for a few minutes, and then the pilot reached forward with the sub's mechanical arm and broke off the top of the pipe. The smoke came out much faster. The pilot then slipped a temperature probe into the thunder spout. Back on the surface, they found the temperature probe had melted. On later dives, their recalibrated probe measured temperatures as high as 662°F, hot enough to liquefy lead and burn out the sub's view ports. From that day forward, scientists have regarded black smokers as if they were fire-breathing dragons from another world, waiting to eat them alive.

At about the same time, space scientists made a major discovery at the Jet Propulsion Laboratory in Pasadena, California. For more than two years inside the massive cubical building where NASA controls its missions to the outer solar system, they had been tracking the progress of a spacecraft called Voyager 2

*After Voyager 2, NASA's Galileo probe (above) revealed still more of Jupiter's splendor.*

as it flew 470 million miles to Jupiter. The vehicle was as big as a living room, weighed almost a ton, and had millions of separate parts. Too far from the sun to be powered by solar energy, it relied on a capsule of radioactive plutonium to energize its computers, cameras, and measuring instruments.

For weeks the scientists had watched as *Voyager 2* approached the biggest planet in the solar system. Slowly, its cameras turned the moons of Jupiter— four tiny dots discovered by Galileo in 1610—into spheres that grew into orbs. On the morning of July 9, 1979, Voyager began sending the first close-up pictures of the smallest Galilean moon, called Europa.

The members of the imaging team and their assistants jammed into a dimly lit room staring at a screen as the images—a series of vertical strips—were transmitted from the spacecraft. After 300 years of looking through telescopes and developing theories, the general consensus was that they would see a small, dead, heavily cratered planet. Instead, they were confronted with a smooth, spherical surface of ice crisscrossed with lines. The absence of craters indicated they were looking at a young, active world. The room of scientists fell silent, followed by gasps of incomprehension. According to Professor David Grinspoon, "Never before and never since has our knowledge of another world taken such a great leap."

Several of the young scientists sitting at the table talking to Cameron had just been born when these dual discoveries were made. They grew up in the 1980s, when America's space shuttle inaugurated a new era in space travel and Soviet cosmonauts were setting records for days spent in orbit. They were enchanted by the sea-space expeditions they saw on television and read about in magazines. In high school, they found that physics and chemistry were cool. They went to college wanting to pursue science as a career.

Because he was older, Cameron surfed an earlier wave of interest in sea-space exploration. He was 15 when Neil Armstrong and Buzz Aldrin strolled across the moon. He was 17 when the United States launched Pioneer 10 toward Jupiter and interstellar space. Cameron studied the Pioneer plaque showing the pulsar map and human figures designed to introduce Earth's civilization to alien life and wondered what cosmic trajectory the spacecraft would take. At the same time he read articles in *Scientific American* and NATIONAL GEOGRAPHIC about undersea projects called *Man-in-Sea* and *SeaLab.* Among those who most influenced his passion for the exploration of inner and outer space were Jacques Cousteau and Carl Sagan.

Cousteau, the acclaimed undersea explorer, co-invented the Aqua Lung, wrote books entitled *The Silent World* and *World Without Sun,* and hosted a television series called *The Undersea World of Jacques Cousteau.* One of the things Cameron found intriguing about the "grand seigneur" of diving was his interest in the technology of exploration. Cousteau designed a diving saucer to carry two people to 1,000 feet. A few years later, he built the undersea stations for his Conshelf project that allowed divers to dwell for weeks at depths as deep as 328 feet. The implication was clear: If you want to go deep into the ocean, you'd better think hard about the tools that will take you there.

Cameron was enchanted with *Le Monde Du Silence*, Cousteau's Oscar-winning documentary

and the television series that followed it. They featured beautiful shots of the wide ocean and the good ship *Calypso*. In sunlit scenes, divers holding phosphorus torches swam through blue depths toward coral reefs and colorful life forms. The shows told fascinating stories about men who swam into an alien world and found themselves at home with squid, salmon, and sharks.

Cousteau repeated this theme in all of his films and television documentaries: the awe divers feel when they see the magical life beneath the sea for the first time. For Jim Cameron, Jacques-Yves Cousteau was a 20th-century synthesizer who embraced art and science and fused them into fascinating films.

Carl Sagan had the same passion for outer space that Cousteau had for the ocean. The professor of astronomy and space science and director of the Laboratory for Planetary Studies at Cornell University played a leading role in the Mariner, Viking and Voyager missions. He published hundreds of scientific and popular articles and won NASA medals for exceptional scientific achievement and distinguished public service. He was the author of books like *The Cosmic Connection* and *Murmurs of Earth*, for which he was awarded the Pulitzer Prize for literature in 1978.

In the 1970s, Sagan used his charismatic communication skills to promote his belief that the search for extraterrestrial life, or exobiology, should be the focus of space exploration. He advocated the search for life beyond Earth with the fervor of a missionary. A 1971 article in *Time* magazine described him as "exobiology's most energetic and articulate spokesman."

The 1960s and '70s were tough years for exobiology, however. When Mariner 4 orbited around Mars, its cameras revealed a lunarlike surface riddled with craters—a big disappointment for a public anticipating a planet with more Earthlike features.

NASA hoped that the follow-on Viking missions would reveal more. The technical heart of the missions was a 30-pound, self-contained biology lab the size of a microwave oven. Inside were 40,000 parts, including sample chambers, canisters of radioactive gases, nutrient containers, grow lights, and Geiger

OPPOSITE: *Callisto's cratered surface may hold a slurry ice-ocean, making it another candidate for extraterrestrial life.*

counters. However, at both Viking lander sites, separated by more than 4,000 miles, the four biological experiments failed to find any signs of organic material. Neither did the spacecraft's cameras. Once again, failure to find life on Mars was a big, public disappointment.

One of Carl Sagan's great contributions was to introduce the world to the mysteries of the cosmic ocean just as interest in life-seeking space probes was diminishing. Sagan and his partner B. Gentry Lee, the Viking mission planning director, formed a production company devoted to communicating science in an engaging way. Working with PBS, they spent three years producing *Cosmos*, a television series focused on astronomy and its influence on human history. The series dramatically raised public awareness of astronomy, and helped to popularize Sagan's belief that we are not alone in this vast universe. As Sagan wrote,

> *…is there anyone out there to talk to? With a third or half a trillion stars in our Milky Way Galaxy alone, could ours be the only one accompanied by an inhabited planet? How much more likely is it that technical civilizations are a cosmic commonplace,*

*that the Galaxy is pulsing and humming with advanced societies…perhaps when we look up at the sky at night, near one of those faint pin-points of light is a world on which someone quite different from us is then glancing idly at a star we call the Sun.*

The television series reached more than 140 million people, and among them was 26-year-old Jim Cameron. "*Cosmos* really hit a chord with me," he says, "The images and music were beautifully composed. So were the artists' conceptions of what outer space looked like. It made me want to jump on a rocket and go there."

At the end of the eighties, NASA returned to Europa with a hard-luck spacecraft called Galileo. The $1.4 billion vehicle was supposed to be launched from a space shuttle in 1982, but technical problems, and the loss of the *Challenger* in 1986, pushed the date back to October 1989. It was programmed to take more than 50,000 pictures during its voyage, but its high-gain antenna failed to open. A small, low-gain antenna was pressed into service by an innovative, never-give-up ground crew. When Galileo arrived at Jupiter to begin its gravity-driven tour in 1995, everyone associated with the project had fingernails bitten to the quick.

Then the images began to trickle in: the towering thunderstorms, Great Red Spot, and shimmering auroras of the king of planets; the exploding volcanoes of Io; the densely grooved surface of Ganymede; the ancient, cratered terrain of Callisto; and the blue and white ice-covered surface of Europa.

Unlike the Voyager flyby missions, Galileo stayed in orbit. For eight years it revolved around the huge planet taking close-up pictures of the Jovian moons. Because the craft carried the first digital camera flown in space, the images came back to Earth with uncanny detail and sharpness. Coupled with data, they revealed that the four biggest moons had the geological complexity of planets. Galileo was exploring a planetary system far more active than anyone thought.

The reason was tidal energy, the same kind of rhythmic force that causes the Earth's moon to lift and release water each time it passes over the ocean. Jupiter, with a gravity two and a half times stronger than Earth's, is pushing and pulling the interior of its orbiting moons, generating a furious heat that drives volcanoes on Io and cracks the icy surface of Europa. This constant flexing keeps their insides hot and their surfaces in motion. As expected, the highest internal heat is found in the moons closest to the great planet, where the gravitational pull is strongest. Innermost Io is perhaps the most volcanically active body in the solar system. Outermost Callisto is heavily cratered, suggesting its surface has been cold and dead for millions of years. In between lies Europa, with an icy surface that appears to be constantly reworked from below. Some planetary scientists think that Europa might be the "Goldilocks Moon": not too hot, not too cold, but just right for supporting life.

During the 1970s and '80s, I had the good fortune to lead 15 expeditions to explore the world hidden under the ice of the Arctic Ocean. This meant flying over hundreds of miles of icebergs, and smooth and shattered ice pushed into pressure ridges. The surface of Europa reminds me of those icebergs and pressure ridges. The tens of thousands of them laced into intricate patterns suggest enormous forces at play.

Some scientists believe that Europa's bright surface is composed of ice crisscrossed with cracks and fissures formed as its surface moves with the tides. Its icebergs seem to have broken loose and then refrozen. Magnetic field measurements indicate that a layer of electrically conductive material—possibly a volume of salt water larger than all of Earth's oceans combined—lies beneath the shifting ice.

# BEAUTIFUL FLUID: WATER AND LIFE

Water is the blood of the Earth. It moves from clouds to streams to rivers to the sea and back again, weathering and shaping everything it touches. It erodes mountains, enhances plate tectonics, governs global climate, and eases the thirst of living creatures. Some planetary scientists believe that in the early solar system water was everywhere, but only one planet, Earth, had the correct size and right distance from the sun to ensure the capacity to hang onto it.

Everyone knows that water is made of two atoms of hydrogen tightly bound to an atom of oxygen. But it is the dynamics within the linkage of this hydrogen bond that makes water "the miracle molecule." Hydrogen and oxygen are powerfully attracted to each other. Every oxygen atom has six electrons orbiting its nucleus, but has room for two more. Two hydrogen atoms provide the missing electrons, resulting in a four-cornered molecule called a tetrahedron. The arrangement of the hydrogen and oxygen gives one side of the water molecule a positive electrical charge and the other side a negative electrical charge.

This electrical polarization has profound consequences. It attracts other water molecules and binds them together with hydrogen bonds. These bonds allow water to exist as a fluid. They are the invisible, quivering force holding together an ocean that covers seven-tenths of the surface of the Earth.

A second consequence of a water molecule's electrically polarized design is that it reaches out and attaches to other things. Water is the universal solvent, bringing other atoms and molecules together and tearing them apart. It provides the essential context for all biochemical reactions.

Water has another life-enhancing property. When it freezes into a solid, it becomes lighter. Most materials become denser when they are solid. The profound consequence is that ponds, lakes, and oceans don't freeze from the bottom up, killing everything within them. If ice sank every time the temperature dropped below zero, we would have long ago said good-bye to all the multicellular organisms living in water.

The miracle molecule is what allows the chemical evolution of simple organic matter into living organisms. Recent discoveries suggest that unicellular organisms living deep within the early ocean may have been Earth's first inhabitants.

The ocean, which composes about 97 percent of the planet's biosphere, is an ideal environment for the emergence and maintenance of life. Its enormous blue surface is bathed in sunlight—photosynthetic energy—from our home star. Its floor is studded with volcanic vents pouring hydrothermal vent fluids—chemosynthetic energy—into its depths. Best of all, it contains 320 million cubic miles of a stable, life-enhancing medium called water.

The first space missions to Mars were designed to search for biological signs of life. After thinking hard about the detection capabilities of systems they were sending into space and what they should be looking for, planetary scientists decided they had to follow a more elusive trail. From Spirit to Opportunity to Cassini, many of today's space probes are designed to "follow the water."

OPPOSITE: *The magnificent combination of oxygen and hydrogen in water is essential to life on Earth.*

A thought occurs to me as I look out the porthole of the *Keldysh*. The ocean below contains millions of species from bacteria to blue marlin. Biologists are now asking the question: Does any life dwell in the depths of Europa's possible polar seas?

Europa has become the most promising place within our system of nine planets to search for water-based life. Sometime within the next decade, NASA plans to launch a mission called JIMO (Jupiter Icy Moons Orbiter) to orbit Europa and confirm the theory of a fluid ocean. The orbiter will also examine the icy surface for the chemicals needed for life. If an ocean is found, a follow-up mission will land on the surface, deploy a self-disinfecting probe to melt a circular hole, and go ice diving in the solar system's deepest ocean.

Everyone around the table has considered the moons of Jupiter and what they might reveal about extraterrestrial life. It is a subject that Hand is studying as a scientist and Cameron is exploring as a filmmaker.

"So Kevin, let's say we go to Europa and put a lander on the surface. It bores down through the ice, pops out into this alien hydrosphere, and goes to the bottom. Are we going to see tube worms? Are we going to see big animals? What's possible?

Hand answers quickly. "I don't think we're going to see any animals."

Cameron smiles. "Wait a minute. That's pretty negative. You sound like the guys who said that the floor of Earth's ocean was sterile. Then, in the seventies, they discovered the vents and their thriving colonies."

Pan Conrad, an astrobiologist from the Jet Propulsion Laboratory (JPL), speaks up. "If there's an ocean, there's water, and if there's water, there's oxygen. That's what's so compelling about Europa."

"Just because it doesn't exist on Earth," says Cameron, "doesn't mean there isn't a mechanism in Europa's ocean that might free the oxygen in the water column. What if there was an organism that could electrolyze the water, using one form of energy to generate oxygen?"

"We've got that," says Conrad. "That's what organisms do. Electron transfer is a game they love."

A round of coffee is poured. "You mean," says Hand, "you want tube worms and squid in Europa?"

"I want stuff you can see, because it's more cinematic. I want intelligent squid with cities built out of ice blocks upside down on the other side of the ice shelf." Eyes blink; Cameron the undersea explorer has morphed into Cameron the Hollywood film director.

Conrad tries to steer the conversation back toward hard science. "I think statistically you're out of luck."

"How do you know?"

"If we don't go look, we'll never know," she says with a grin, "So let's go look."

A glance out the porthole across the fiercely bright water sets the mind to thinking that the men and women sitting at the table are imbued with the restlessness and ambition of the pioneers who went into the American West. Those stout hearts crossed wide plains and climbed over mountain ranges. These stout hearts cross wide seas and dive down to mountain ranges. Both explore vacant and monumental spaces where what is human is small and lonely.

Like all disciplined scientists, everyone around the table started a long time ago with hard questions. How did Earth's ocean get here and how did it bring life into being? Why have recent discoveries about

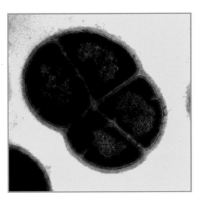

*Evidence of the resiliency of microbes, the extremophile* Deinococcus radiodurans *can survive 1,000 times the radiation of a nuclear blast.*

the resilience of microbes changed the thinking about life in other parts of the solar system? What is the possibility of life on Mars? These questions have been provoked by recent developments in scientific thought.

We know that 97 percent of all the water on Earth—320 million cubic miles—is contained in the oceans. Most of it is in the Pacific, which covers nearly half the planet and is big enough to swallow all the continents.

In the past 30 years, theories about how the ocean came to dominate the Earth have changed dramatically. Many scientists now believe the planet was formed from the uncounted collisions of huge chunks of rock called planetoids. When the embryonic Earth was the size of Mercury, its gravity accelerated incoming rocks to a speed that shattered and melted them on impact, releasing the water in their interior. From its violent, Hadean beginnings, the Earth formed wet.

This idea is augmented by another provocative theory. Beyond the orbit of Pluto—in a spherical shell that surrounds the entire solar system—are a

thousand billion comets whose long, looping orbits sometimes bring them into Earth's corner of the solar system. Comets are huge accretions of dirty ice and frozen carbon dioxide. Some scientists believe that comets bombarded the Earth for hundreds of millions of years "carrying the ocean on their backs."

In this scenario, a cloud of scalding vapor lay over the early Earth like a miles-thick blanket. The weight of the high-pressure atmosphere was roughly a million trillion tons. Eventually the stream of incoming projectiles tapered off and the Earth began to cool. Time passed, torrential rains began to fall, and the water rose in an ocean that was boiling hot.

At some point early in its history, a planetoid the size of Mars struck the Earth. As it punched through the atmosphere, it created a fireball of 60,000 Kelvins, or ten times the surface temperature of the sun. Its impact set off a series of earth-shaking seismic and volcanic events. The incoming planetoid was mostly vaporized, but some of it rebounded back into space and formed the solid body we call the moon.

Until recently, conventional wisdom held that life began at the surface of the ocean in Charles Darwin's "warm little tide pool." Today, many biologists believe that life may have originated far below the surface at hot springs on the floor of the ocean.

When the early Earth was molten rock and the "ocean" was emerging as a globe-circling cloud of superheated water, it was a place unsuitable for life as we know it. Then, 3.8 billion years ago, the rain of planetoids ended and conditions began to stabilize.

In ancient rocks from western Australia, scientists have discovered 3.2-billion-year-old microbes that must have lived through the bombardment of Earth by comets and planetoids that marked its first billion or so years. Some of these cellular filaments resemble bacteria and cyanobacteria or blue-green algae. The discovery suggests that heat-loving organisms living in the depths of the early ocean may have been Earth's first inhabitants. Their probable habitat was near the hot water flowing out of deep-sea volcanic vents.

The how and where of life's origins on Earth is

OPPOSITE: *The flaky composition of this Mars rock, dubbed "Mimi" by NASA scientists, may be caused by the action of water, providing another clue in the search for extraterrestrial life.*

still a matter of speculation, but one thing is certain. Life is a dynamic, self-organizing process that requires water and energy. In our solar system, Earth is the only planet with liquid water on its surface.

During more than three billion years on Earth, microbes such as bacteria and archaea have evolved to occupy every square inch of the surface of the planet. Examples of this microcosmos of creatures hidden from human eyes are found everywhere from the top of Mount Everest to the depths of the Mariana Trench. Antarctica's McMurdo Dry Valleys area is one of the coldest, driest, and least nutritional places in the world, but its soils are home to a host of photosynthetic bacteria, single-celled algae, and microscopic invertebrates feeding on these primary producers. Polar sea ice, which covers millions of square miles of the Arctic and Antarctic Oceans, does not seem a likely place for life to thrive. However, its interior is riddled with brine channels in which single-celled algae flourish year-round.

Extremophiles are specialized microbes—superbugs—that have adapted to extremes of heat, cold, pressure, and/or radioactivity. The superlatives are astounding. A bacterium called *Pyrolobus fumarii*, found at hydrothermal vents, can reproduce at 235°F. In 1996, Japanese scientists dropped a remotely operated vehicle seven miles into the Pacific Ocean and recovered samples containing hundreds of species of archaea, bacteria, and fungi. The bacterium *Deinococcus radiodurans* can survive a thousand times the lethal radiation of the atomic explosion at Hiroshima. Scientists have found hints that the hardiest of the superbugs might be carried aloft into the stratosphere, drift away from the Earth on the solar wind, and land on the surface of other planets such as Mars.

Recently, enormous assemblies of bacteria and fungi have been discovered living in the pores of igneous rocks miles deep under the Earth. These buried microbial ecosystems, independent of life on the surface, obtain their energy from inorganic chemicals within the rocks. They confirm the premise that whenever there is liquid water, organic molecules, and a source of energy, there is life.

Mars is a bitter-cold desert with tattered clouds, dust storms, reddish dunes, and a sky the color of warm butterscotch. The only visible

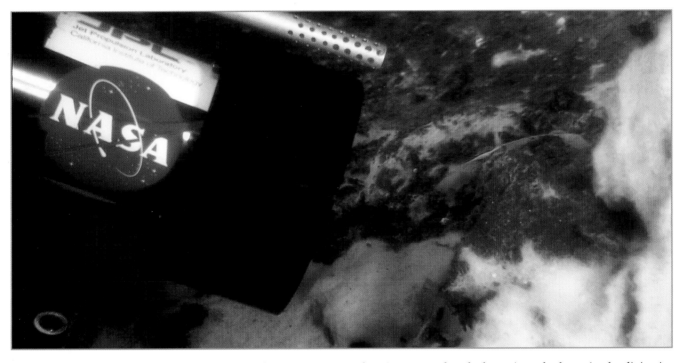

*"Mats" of bacteria are common at hydrothermal vent sites. Recent theories suggest that the bacteria and other microbes living in upper layers of the Earth's crust may weigh more than all other living things on the planet.*

evidence of water has been found at the north and south poles, where there is a frigid mixture of carbon dioxide and ice. However, many planetary scientists believe that in its early days Mars was much warmer and wetter.

The Mars Global Surveyor, a spacecraft that went into orbit around the red planet in 1997, took photographs and altitude measurements that suggest ancient shorelines and wave-pounded terraces running parallel to these shorelines. Among them are huge flat areas implying a central basin with thick accumulations of sediment.

These are tantalizing signs of what might have been an ancient Martian ocean existing more than a billion years ago in the northern lowlands. It could have

# ASTROBIOLOGY AND ALIENS

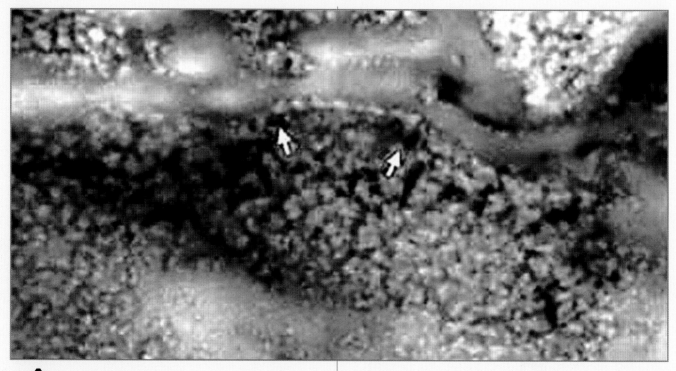

Astrobiology—the combination of astronomy and biology—is a new word describing an old idea. As far back as the 17th and 18th centuries, thinkers like Galileo and Huygens gazed at the plenitude of stars and believed that the universe was a vast and fertile place, and possibly inhabited by other creatures. It is a belief that has waxed and waned ever since.

In the mid-1990s, the possibility that life might exist beyond Earth got a boost from three events: the discovery of the first planets outside our solar system, the possibility of fossils in a meteorite from Mars, and images from the Galileo orbiter indicating the likelihood of an ocean on Europa.

In 1995, astronomers discovered a planet orbiting around a star called 51 Pegas. Soon after, dozens more were found in other parts of the sky. To date, more than a hundred have made the list. In 1996 at a press conference in Washington, President Clinton announced that scientists had found organic molecules, chemical traces, and possible microfossils in a four-pound Mars rock that landed in Antarctica in 1984. The "aliens from Mars" made one-inch headlines around the world. Astrobiology became the new buzzword within NASA and the space science community.

In 1998, the Astrobiology Institute was established at NASA's Ames facility in California's Silicon Valley. Its budget was 5 million dollars. Four years later, its budget had risen to 15 million dollars. Centers affiliated with the Institute sprang up in five other countries including France and Spain. The objective is to bring astronomers, geologists, biologists, and planetary scientists together, break down the barriers between their specialties, and exchange hard-won ideas about the why and where of life.

So what would an "alien" look like? "In a word, microbes," says Pan Conrad, an astrobiologist at NASA's Jet Propulsion Laboratory (JPL) in Pasadena, California. "On Earth, most of the biomass is microbial, and most species of life are microbial, so by volume and numbers, microbes rule."

Extremophiles or "superbugs" are microbes that have adapted to physical and chemical extremes. They have been found everywhere on Earth, from deep within the crust to the top of Mount Everest. "Microbes can live in a wider range of environments than any other form of life on our planet, says Conrad. "They can withstand more cold, more heat, more pressures, and more chemical extremes. They can breathe air, reduce metals, and make food out of light. They are just very clever little buggers at finding ways of doing business."

Tori Hoehler, an astrobiologist at NASA's Astrobiology Institute, agrees. "The prevailing wisdom suggests that if life exists on Europa or other places in the solar system," he says, "it will likely be microbial life. Whatever form it's in, it would help answer one of the most important questions ever asked. It would be a profound discovery."

Until future space probes land on faraway orbiting shores, scientists like Conrad and Hoehler will continue exploring provocative places like the sunless floor of the sea.

*Compelling evidence of possible signs of life on Mars literally fell to Earth in 2001 in the meteorite at left. Some scientists believe that the spots indicated by arrows show ancient bacteria.*

had long beaches, waves, and hidden depths. It would have covered about one-sixth of the surface of the planet—an area bigger than the Mediterranean and Arctic Ocean combined—with water up to a mile deep.

That Mars might be a second oasis in our solar system inspired a new scientific strategy: Follow the water. In the summer of 2003 an international caravan of robotic vehicles made its way toward a rendezvous with Mars. Among them were the European Space Agency's Mars Express, the Japanese spacecraft Nozomi, and the two NASA rovers Spirit and Opportunity.

Inevitably, the conversation turns to Mars and how to explore its faraway surface. Cameron starts off by sharing his thoughts about the *Mir* subs and what they might tell us about traversing the surface of Mars.

"Let's say a *Mir* dive is the equivalent of a pressurized rover going out on the Martian landscape," he says, "Look what it takes: 20 guys to get the sub ready. Then there is the launch and recovery, nine hours on the bottom and the acquisition of the samples. After the dive, the samples have to be recorded and stored. It takes an entire day."

"You won't have 20 guys on Mars," says Kelly Snook.

"You won't have 20, not for a while. For the first few missions you'll have five or six people, all of them multitasking individuals who can do all kinds of things. They may not be scientists, but they'll have a good understanding of how science works."

Cameron studies the faces of the three young people sitting across the table. The expedition is giving him new respect for the rigor of their field work. "Now that we've followed you around with our cameras," he says, "we've seen how much you do every day and how hard it is."

The conversation turns to the differences between using robots and using humans for exploration. Cameron turns to Kevin Hand and asks, "Would you go to Mars if you had the chance?"

"Sure," says Hand with a smile.

"What if they said we need an astrobiologist to go to Mars, and you're going to have to train for ten years, and it's going to be two and a half years of your life, and the probability of your coming back to Earth is maybe 80 percent?"

The smile starts to slide from Hand's face. "I'd still go," he says.

Cameron knows that space exploration can be the hardest work anyone can love. As a member of an advisory panel to NASA, he's talked to astronauts who have trained for years and have never set foot on the International Space Station. Some suspect they may never get there. He wants to find out how determined Hand really is.

"To find a bacterium?" he says softly. "So, if you get really lucky, you can say, 'Hey look, we found a bacterium?'"

Hand speaks slowly, measuring his words. "Of course I would go. I would be motivated by the 80 percent explorer in me and the 20 percent scientist." He pauses and looks directly at Cameron. "But the scientist would first ask, 'Do I need to go? What's the reason? Can a robot fulfill my function?'"

Both men are aware that it takes an average of six minutes for a radio signal to travel from the Earth to Mars, which means controlling a robot on Mars from Earth is painstakingly difficult. Cameron says, "What if you could be on Mars, inside a station, doing your science with three or four robots and no time delay? With your machines on the surface right in front of

*Cathedral-like spires at the Lost City vent field are formed by a chemical reaction between seawater and mantle rock, a process unique to this vent field and believed by some scientists to be the likely "look" of vent fields on other planets.*

you, taking samples and making critical measurements?"

There is a long silence. The Hollywood director has morphed back into the polymath who makes presentations to the Mars Society and talks to senior management teams at NASA about exploring the planets in our solar system.

He turns and looks directly at Figueroa. "What about you, Dijanna, would you go to Mars if you thought that there was interesting life there?"

"Sure I'd go. The explorer in me says GO."

"What if the science community thought there wasn't going to be interesting life there?" someone asks.

"I'd still go."

For days on end, the morning sky above the *Ares* and *Keldysh* looks dry, white, and inflammable. There are no waves, only low, smooth-topped swells. Where light strikes the ocean, it looks like it is made of sun.

One of the subjects discussed during the months on the Atlantic and Pacific is the growing evidence that a rapidly changing climate is driving many species into extinction by shrinking or altering their habitat. The UN's Intergovernmental Panel on Climate Change predicts that global temperatures may rise between 2°F and 6°F during the next century. If that happens, between 18 percent and 35 percent of the planet's species could be extinguished by the year 2050. An extinction of this magnitude—possibly the largest since the dinosaurs were killed off 65 million years ago—would have profound effects on the human family by altering the geography of food production and increasing the spread of disease. A worrying thought to all the scientists is that just as we are beginning to understand the origins of life, we are witnessing its rapid extinction.

We are subject to warming of a more immediate sort. There are days when the sun-blasted decks of both ships feel like ovens. The men and women standing next to the subs try to ignore the heat as they exchange thoughts about the upcoming dive and the daily ritual of work. All have jobs that require impeccable teamwork and timing, tasks that force them to share a closeness with the subs and with each other that began the day they boarded the ship.

Their bodies are bent-shouldered and sweat drained. The blistering heat slows all movement. There are no smiles on their faces, just dogged determination. They work in the spirit of the first generation to see the bottom of the sea and the surface of the planets transformed from distant objects into comprehensible places. They toil like those who may be the first to learn that we are not alone among the stars.

In an hour the first of the subs will be lowered into the water and begin its descent. On the way down to the hydrothermal vents it will pass through water teeming with invisible bacteria, archaea, and protozoa—part of a vast, swarming pyramid that extends all the way to the seafloor. Their ancient presence suggests that if we travel far enough into the sea and out toward the stars, we will be moving toward the center of our existence.

OPPOSITE: *Fiery walls of compressed gas in the Bug Nebula conceal one of the hottest stars known in the universe. Might it also conceal planets with the conditions right for the support of life?*

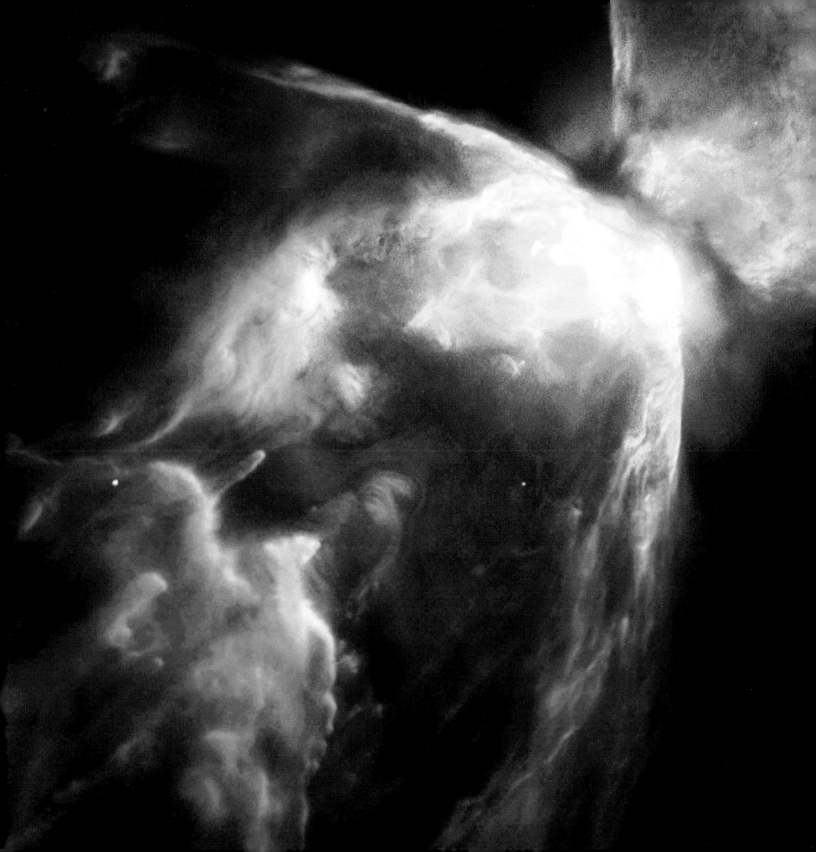

Everyone who reads science fiction understands the concept of a point-and-shoot life detector," explains Pan Conrad. "We're trying to come up with that device." The NASA/JPL device pictured here uses invisible laser light to illuminate samples and "read" their fluorescence. Certain organic compounds have signatures that indicate very clearly that something is biological. Here the device takes a reading from a field of sulfur-oxidizing bacteria that the science team nicknamed "sunny-side up slope."

Breathtaking as a Himalayan vista, the spires of Lost City are illuminated in ghostly blue by the *Aliens* team. Scientists recently discovered that, unlike other vent systems, Lost City may be more than 30,000 years old.

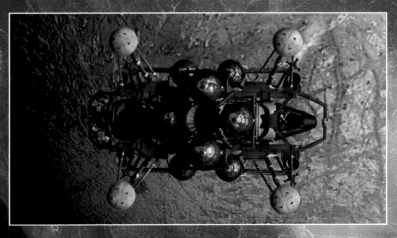

Cameron's imagined future mission to Jupiter would land on Europa's icy surface, then drill through the ice (ABOVE), to finally explore an alien ocean (LEFT).

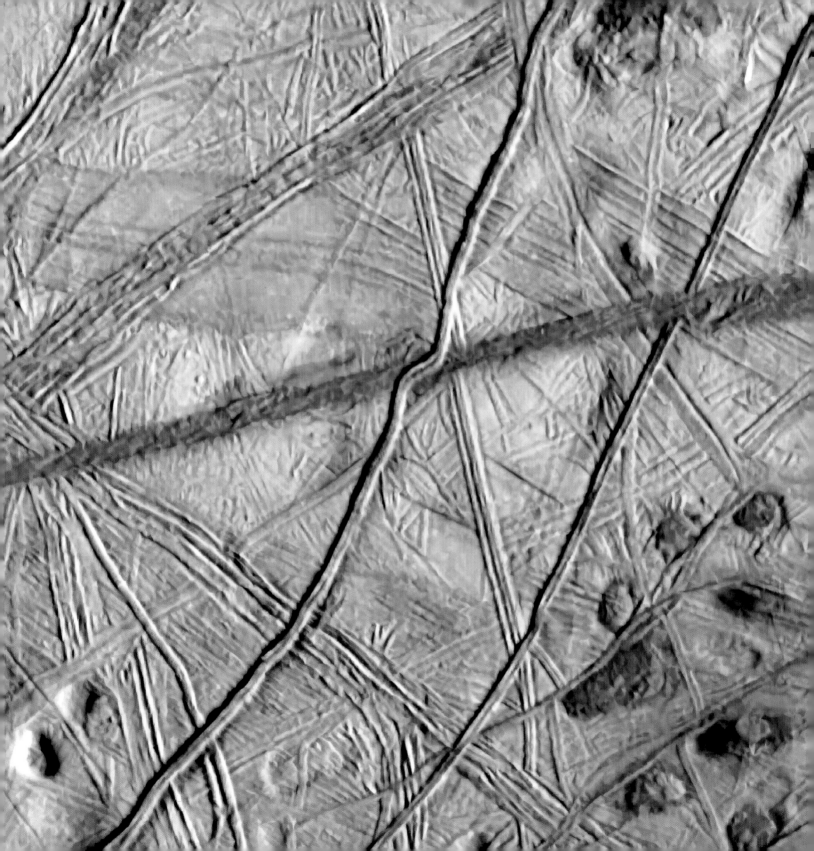

The size and spacing of the marks that criss-cross Europa's surface suggest that the moon's icy shell may be in motion, with warmer ice moving upward and colder ice sinking down. Other evidence hints that a deep melted ocean may lie beneath the ice. The ice erupting onto the surface may hold clues about the ocean's composition.

# CHAPTER FOUR

# THE PACIFIC DIVES

### INSIDE THE DARKNESS

Nicknamed "paratrooper crabs" due to their tendency to leap off things and drift down, a pair of Galatheid crabs stands atop a Pacific outcropping.

On September 20, under an archipelago of white clouds, the *Akademik Keldysh* steams out of the western end of the Panama Canal and heads into the world's largest ocean. The waters shimmering before her run for 11,000 miles until they reach the shores of Asia.

*The rusty color of the* Riftia *observed at 21ºNorth indicates that this vent site may be dying.*

At a speed of ten knots, the big white ship takes three days to reach the first dive site. Her heading is north by northwest over a stretch of tectonic seafloor known as the Cocos Plate. Far to the south is the Galápagos vent site whose discovery stunned the scientific world in 1977.

Among the scientists on board for this leg of the expedition is Jim Childress, a professor at the University of California at Santa Barbara. In 1979 Childress was a principal investigator on the first biological expedition to the Galápagos site. He and his coworkers used the research sub *Alvin* to observe the vent animals and collect the first samples. The savage, swarming life in the bitter black water turbocharged his career. For the past 25 years, he has been studying the chemical symbioses of animals in the East Pacific vent fields. He and his team have developed techniques for bringing up live specimens, maintaining them at high pressure, and studying their hydrogen-sulfide-based metabolism. Dijanna Figueroa is one of his graduate students. Joining Childress and the other scientists on the *Keldysh* are Pan Conrad, astronomer-planetary scientist Arthur "Lonne" Lane, Tori Hoehler, and marine physiologist Michael Henry.

During the next month, under Jim Cameron's direction, the *Mirs* will make three double dives at vent sites 9ºNorth, 21ºNorth, and Guaymas. The 9ºNorth vent field (nine degrees north of the Equator) is the only active vent field studied shortly after an underwater seismic event decimated its community of animals. The 21ºNorth site is where black smokers—mineral

*While not specifically dependent on the energy of the vents, anemones tend to propagate in these environments.*

chimneys spewing thick clouds of hydrogen sulfide—were first discovered in 1979. The third vent field is at 27° north latitude in the Guaymas Basin, halfway up the Gulf of California.

The *Keldysh* is lying in the swells, her engines idling, about a thousand miles northwest of the Galápagos Islands. Some 8,250 feet beneath her keel is the East Pacific Rise at latitude 9°50' N. Within the Rise is an eruptive fissure or axial valley containing a linear array of black smokers and low-temperature vents.

In April 1991, scientists in *Alvin* saw the aftermath of a volcanic eruption at this site. A portion of the ridge was layered with fresh basalt. The ocean floor was covered with dead and dying tube worms and mussels. Shreds of bacterial matter were being blown "like snow" out of holes in the rocks. The shreds were so thick that the sub pilots were blinded and had to rely on sonar for guidance. Everything was covered in gray ash. Deep-sea biologist Rich Lutz of Rutgers University and his colleagues called it "the tube worm barbecue" site.

No one is certain what caused the eruption.

One theory is that the magma beneath the site shifted upward and displaced the water in the cracks of the crust. The pressure built up until the cap of magma blew apart, sending shards of basalt in every direction. Hydrothermal minerals spewing from the seafloor fell as a rain of gray ash.

Whatever happened created a stage zero in the life cycle of a vent. On return visits in 1992, '93, and '94, scientists documented a phoenix-like rebirth of the colorful cornucopia of life. Nine North became a place where scientists could see how fast black smoker chimneys grew and how quickly animals recolonized them.

Three hours ago, on the deck of the *Keldysh,* Mike Cameron climbed down into the pressure sphere of *Mir Two* and started to run through his predive checklist. He tested his robot's small camera and lights. He spun its thrusters and extended the new manipulator that he had installed after the Atlantic dives. The launch over the side of the big Russian ship is as smooth as a slug on a salt lick. During the descent, he talks quietly to Victor the pilot and biologist Loretta Hidalgo. On a previous dive they found the site was still cloudy, but spotted many bacterial mats, octopus,

*Resembling a crumpled-up piece of paper, this lava field at 9ºNorth was formed by fast cooling of very hot lava.*

anglerfish, flying sea cucumber, and healthy tube worms. Some "tumble-weed stuff" was seen that had crabs inside it.

After two hours of freefall, the seafloor beneath the sub is revealed in a cone of light and comes up like a charcoal dawn in a volcanic wilderness. After 40 minutes they rendezvous with *Mir One.* As soon as the two subs are in position, they get down to business. They film the glassy surface of fresh lava flows glaring like ice. They fly over vertical rock walls where crustal movements have broken and displaced lava flows.

At one point, they stop and deploy a cylindrical instrument from the Jet Propulsion Laboratory called a fluorometer, an instrument that measures the presence of biological compounds.

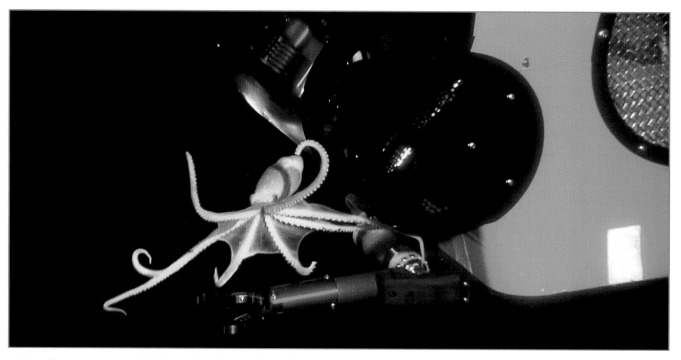

*In a close encounter with the team's ROV. a minuscule vent octopus takes a turn investigating the film crew. Several species of octopus inhabit hydrothermal vent communities, preying on mollusks found at the vents.*

As they lift off and move across the tortured terrain, they see high, hot pleats of shimmering water and "lakes" of lava with their surfaces frozen in marvelous swirls. They are able to film an unidentified feathery creature, *Riftia*, and mushroom vents.

In places, the molten lava has drained away from a lake, leaving a skin of basalt as thin as the glass dome over the *Titanic's* grand staircase. As the lava receded, the basalt collapsed into sharp-edged rubble on the floor of the lake. Staggered lines of lava pillars mark the edge of the lake.

Toward the end of the dive, they park downslope from a ridge of broken basalt. It is here, almost two miles under the Pacific Ocean that

Mike Cameron has an out-of-body experience. He knows he is crouched inside *Mir Two*. He is aware that *Mir One* is a short distance away. But most of his imagining brain is inside *Jake*, the small blue robot 50 feet in front of him, slowly climbing up the ridge. A voice tumbles out of the sub's intercom. "Come right. Come right about 35 degrees, it's above you." The words are from his brother Jim in *Mir One*.

So far on this project, Mike has made nine dives in the Atlantic and Pacific Oceans. From his cramped position inside *Mir Two* he has operated *Jake* at four vent sites, including Menez Gwen, Lost City, and Snake Pit.

Mike's thumb and forefinger are curled around a small joystick in his lap. His eyes are glued to a flickering video screen. He eases the joystick back, sending a stream of electrons pulsing through a hair-thin fiber-optic cable that leads out of the sub and across the seafloor. *Jake* tilts up. Without warning, the center of the video screen fills with an octopus as white as a ghost orchid.

Mike Cameron has been scuba diving for more than two decades. He's operated his robots as deep as 16,000 feet and had encounters with hundreds of animals, from invertebrates to mollusks to mammals. Even with all this experience, he's not prepared for what happens next.

The octopus is poised on the rimrock of the ridge. Instead of fleeing into the shadows, it begins to advance. It creeps forward, sliding out one tentacle and then another, pulling itself over the rocks, pausing briefly as it draws in the tentacles behind it.

Cameron—*Jake*—slows his forward motion. Aware that his brother Jim is filming this encounter, he rotates and slips sideways to stay within the frame of the 3-D-HD camera. He reduces his speed so that he is less threatening. Cautiously he extends his short blue metallic arm. Then he descends until he is hovering just above the seafloor at eye level with the curious creature.

It has eight tapered tentacles trailing out from a smooth, round body of unblemished whiteness. Each tentacle is festooned with rows of circular pink suckers. High on the body are the dim outlines of a pair of dusky ovals. These "eyes" are locked onto *Jake*.

For an instant, Mike Cameron wonders why an animal living in a place of eternal blackness needs eyes. Do they detect bioluminescence? Are they sensitive to infrared? Whatever their function, they are staring in his direction.

The octopus unfurls its forward tentacles and continues slowly down the slope. It moves unhindered, like water slipping through water.

To hold his position in the current and keep the animal in the frame and in front of him, Cameron applies every skill he's learned as a helicopter pilot and robot pilot. It's one thing to sit inside a control van on the deck of a research ship, drinking coffee while you steer a big robot through the water thousands of feet below you. It's quite another to be hunched inside a sub with almost two miles of ocean over your head and a weird, white creature heading in your direction.

The view ports of both subs are filled with faces looking at the two of them—*Jake* and the octopus. One is a bundle of biochemical reactions within an elastic skin. The other is a collection of carbon composites and circuit boards. Then the octopus reaches out with two tentacles and wraps them gently around *Jake*'s mechanical arm.

For a few fleeting seconds, the octopus tightens its grip and tries to close the distance between itself and *Jake*. Then it releases its hold, pulls in its tentacles, and retreats. As it makes its way back toward the rimrock, Cameron experiences a strange sense of loss.

Long after the adrenaline stops flowing, he will say, "I felt like I was on another planet, shaking hands with an alien."

A few days later, at the 9°North site, two women wearing sky-blue coveralls step through a steel door on the starboard side of the *Akademik Keldysh*, pause in front of *Mir Two*, and talk briefly about the dive they are about to make. Nearby, Russian technicians check the sub's manipulators and scientific payload. The air is filled with voices and the smell of diesel exhaust. The green-painted deck beneath the subs is slick with the glaze of water and hydraulic oil.

The older woman is Maya Tolstoy, 36, a research scientist at Columbia University in New York. She has closely cropped black hair and a radiant, infectious smile. The younger woman is Dijanna Figueroa, 24, a graduate student at the University of California at Santa Barbara. She too has a winning smile.

Both women have spent months at sea on scientific cruises. Each has made long, deep dives in research subs like *Alvin* and *Mir*. But at this moment they are sharing the soft adrenaline swoosh that comes from not knowing precisely what's going to happen when the dive gets under way. They check the items—cameras, notepads, and sweaters—they will take inside the sub, and then they check them again. They discuss the dive plan they heard moments ago at the predive briefing. They glance at the horizon for signs of deteriorating weather. Then they follow their stocky Russian pilot up the tall metal ladder leaning against the *Mir Two*. One after the other, they slip down into its interior.

Once inside, the world suddenly shrinks around them. They are confined inside a space filled with steel panels, switches, glowing lights, meters, and microprocessors. They have little knowledge of how the life-support, communication, and navi-

*Maya Tolstoy readies a monitor to record seismic activity and acoustical data on the seafloor.*

gation systems work, but they trust their Russian pilot. As *Mir Two* is lifted off its cradle and moved across the deck, they listen to the soft hum of the carbon dioxide scrubber and the muffled roar of the heavy-duty launch crane.

It takes almost two hours for the sub to free-fall to the bottom. As gravity pulls it deeper, it passes through sunlit and then sunless layers brimming with marine snow, salps, and siphonophores. Occasionally the women glance out the view port. Mostly they sit quietly talking or looking at their notes.

Tolstoy is a marine seismologist and associate research scientist at Lamont-Doherty Earth Observatory at Columbia University. She uses underwater microphones—hydrophones—placed in the ocean by the National Science Foundation (NSF) and the National Oceanic and Atmospheric Administration (NOAA) to listen for deep-sea earthquakes and volcanic eruptions.

"Our planet is a very dynamic place," she says, "with massive plates many miles thick sliding across its surface, colliding into each other, building mountain ranges and disappearing into the heat of the mantle below. The most dynamic are the mid-ocean ridges where two-thirds of the Earth's surface is formed. Earthquakes are the manifestation of the movement of the plates and the formation of the ocean floor. They can tell you when magma is moving through the crust, about changes in the hydrothermal flow, and when plates are pulling apart. When we study earthquakes, we are imaging the processes deep in the ocean crust where cameras cannot travel."

Once a year, teams of scientists journey out to these hydrophones to download the data and change the batteries. The information helps pinpoint the physical forces shaping the mid-ocean ridges, increasing our understanding of them. This trip to 9°North is part of a three-year instrument-deployment program.

Figueroa is a marine biologist at the Marine Institute of the University of California at Santa Barbara who studies how marine animals—specifically mussels—adapt to their environ-ments. She collects mussels at deep-sea vents and puts them into small, pressurized chambers to maintain the same pressure, temperature, and water chemistry where they were found. On the ship and then in her Santa Barbara lab, she studies their uptake of oxygen and different nutrients. Under the direction of Professor Jim Childress, she is trying to figure out the characteristics, including heat-shock proteins, that distinguish deep-sea vent mussels from their common shallow-water cousins.

Using a lightweight crane on the stern of the *Keldysh,* Tolstoy has already placed her seismic measuring equipment in the water and let it fall to the floor of the ocean. On this dive she is going to make observations about the seafloor where it landed.

"Over the last decade or so," she says, "we have developed tools that tell us when eruptions are occurring along certain parts of the mid-ocean ridge system. We no longer have to rely on the slim chance that we stumble across a just-active area the way we did at 9°North more than a decade ago. We've now observed seismicity associated with seafloor eruptions in the North East Pacific,

*Waiting for fortune—in the form of drifting food—to float by, sea anemones are common inhabitants of vent communities. Their snowy white tentacles catch passing food and feed it into the mouth at the center of their bodies.*

the Arctic, the Mid-Atlantic, the East Pacific Rise, and the Galápagos Spreading Center. We see fascinating differences in the scale of the eruptions at these sites. At slower-spreading sites the crust is colder and it takes more energy for the magma to break through the crust. Generally, the slower-spreading the ridge, the bigger the earthquake."

The seismic equipment Tolstoy has placed at 9°North will monitor the very tiny earthquakes that change the fluid flow through the vents. This might shed light on how the vents are fueled and how these changes affect fluid chemistry and the animals relying on that chemistry.

"If we also record a major eruption," she says, "it

# THE FIRE THAT FRAMES THE PACIFIC

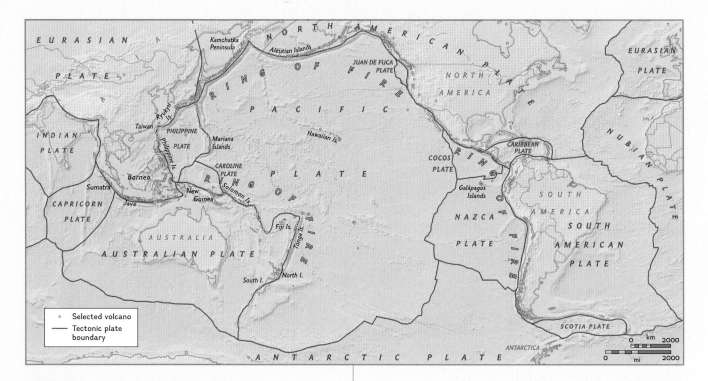

The Ring of Fire is a gigantic geological necklace that runs through and around the rim of the Pacific Ocean. More than 30,000 miles in length, it winds up the west coast of North America to the Aleutian Islands, crosses to Russia, slants down near Japan and the Philippines to New Guinea and New Zealand, and finally heads toward the coast of South and Central America. Because it lies on the moving boundaries of the Pacific Plate, it gives birth to an extravagant number of earthquakes and volcanoes.

A major component of the Ring of Fire is the East Pacific Ridge system (its southern section is called the East Pacific Rise), running roughly parallel to the coasts of South and North America. Like its twin in the Atlantic, it is a raised chain of uplifted crust, a long, fractured wound in the Earth where heat erupts, new crust is formed, and plates separate.

At mid-ocean ranges throughout the world, the seafloor spreads at variable rates. The Mid-Atlantic Ridge moves apart at a rate less than one inch a year and is known as a "slow-spreading" ridge. The East Pacific Rise is "fast-spreading," moving out from its axis at about twice the rate of the Atlantic ridge.

The cross-sectional shape of a ridge varies dramatically with its spreading rate. The slow-spreading Atlantic ridge has a central rift valley that can be three to five miles wide and a mile and a half deep. The fast-spreading East Pacific Rise has a narrow axis that is more depression than valley.

When seen from a distance, all ridge systems, including the East Pacific Rise, have a staggered "staircase" appearance. When these linear segmental boundaries become etched into the sides of the adjoining plate, they are called transform faults and fracture zones.

The magma chambers under a ridge are as unusual as the structures above them. At slow-spreading centers the crust is thick and the chambers are separated from each other. Under fast-spreading centers, the crust tends to be thinner with long, discontinuous chambers pushing up new lava. The structure and movement of the chambers is directly related to the longevity of the vents and their rich biological communities.

Recently formed volcanic crust—fresh, glassy-black lava—is usually free of sediment. However, within a few hundred yards of the upwelling, the lava is weathered and dusted with a slow rain of particles. Within a mile or so of the ridge axis the lava is buried under a carpet of tan-colored sediments.

After the discovery of the Galápagos vent field in 1977, scientists followed the hot-spring trail west to the East Pacific Rise. In the late seventies and eighties they found biological bonanzas at 9°North, 13°North, 21°North, and in the Guaymas Basin in the Gulf of California. In the 1980s, scientists discovered more sites in the north Pacific.

The dramatic distance that separates the Atlantic and Pacific vents may explain why to date, tube worms and clams have been found only in the Pacific. There appear to be limits to the flow of genes between the vents. But time is on their side. There are highways hidden in the ocean, and larval excess in pursuit of eternity is a powerful, driving force.

OPPOSITE: *The notorious Ring of Fire that defines the Pacific Ocean generates millions of earthquakes per year.*

137

would be the ultimate in marine seismic monitoring. Normally, recording devices are hundreds or thousands of miles away from the source. To have our instruments right above an eruption would be as cool (or hot) as it gets in seafloor seismology."

Tolstoy studied geophysics at the University of Edinburgh in Scotland. In the summer between her third and fourth years she traveled to Scripps Institution of Oceanography for an undergraduate program. "They took me on a cruise for ten days," she says, "and I fell in love with the idea of doing science under the sea."

Tolstoy went to graduate school at Scripps, earned a two-year postdoctorate degree, and then moved to Lamont-Doherty. Since 1978, she has participated in 24 cruises and has been the chief or co-chief scientist on 12 of them. "What continues to inspire me," she says, "is the sense of novelty and challenge in the exploration of the seafloor."

Figueroa grew up in Long Beach, California. She had a supportive family, for whom education was a top priority. Her grades 10 and 11 science teacher, Mr. Bradshaw, loved marine biology and often took his class to the edge of the Pacific so they could see tide pools, seaweed, and big white-backed waves breaking on the beach. "He taught us physics and chemistry and how currents moved within the ocean," she says. "He showed the animals and plants that lived in the coastal zone. But his greatest gift was his passion for science and his love of the planet. It was contagious."

In 2001, Figueroa graduated from UCLA with a bachelor of science degree in marine biology. Soon after, she was awarded a five-year doctoral scholars fellowship for postgraduate work at the Marine Science Institute. While in college, she did community service work at the Center for Academic Research Excellence in Los Angeles. An uplifting motivational speaker, she encourages students from socioeconomically disadvantaged backgrounds to pursue careers in the physical and life sciences. "It's my way of giving back," she says, "to all those people like Mr.

OPPOSITE: *The brilliant red of tube worms at Guaymas, originally thought to be a different species from those at other Pacific vent sites, is actually due to an absence of vent crabs. With no crabs to nibble on their ends, the plumes never develop the scar tissue observed elsewhere.*

Bradshaw who helped inspire my love of science."

Within minutes of reaching the seafloor, the two women are looking into the steep shoulders and fuming snouts of black smokers. In the cooler waters nearby, a thicket of red-tipped tube worms leans gracefully into the current. Although trained differently, the scientists are searching for the same thing—significant patterns. Their eyes range over the arc-lamped seafloor picking out objects. Is it animal? Is it rock? Is it big, small, moving, or sessile?

Maya knows that more than a million quakes jolt the Earth every year, and most of them occur under the ocean. Knowing when these eruptions are occurring in real time has led to "response cruises" where scientists quickly steam out to the site to study the aftermath of an eruption. In several cases they were stunned by the amount of bacteria and biomatter floating in the water. Their findings seem to confirm the theory that there is a substantial subsurface biosphere at mid-ocean ridges.

Dijanna eyes the bed of brown mussels beside the tube worms, calculating the movement of water and nutrients into and out of the slit between their calcium shells. She can picture the flow of oxygen and hydrogen sulfide across their gills and the chemical breakdown performed by their endo-symbiotic bacteria.

The fact that the two women are locked up inside the strongbox of a research sub at 8,000 feet is a tribute to their love of science. But it is also a salute to human curiosity concerning the question of cosmic aloneness. Life on Earth is a short burst of flame that blazes briefly in the darkness and then is extinguished. Everything that is good about the human family arises out of the brilliance of this flame.

In the last hour of the final dive of the two-month-long expedition, *Mir One* is 1,000 feet below the surface, heading up toward the late afternoon sunlight on the Gulf of California. *Mir Two* has already been recovered to the deck of the *Keldysh*. Inside *Mir One*, hemmed in by camera equipment, notes, and computers, Jim Cameron and astrobiologist Pan Conrad are leaning back against the curve of the pressure hull as pilot Anatoly Sagalevitch talks in Russian to the communication control team on board the mother ship.

It has been a good dive. They made a soft

landing at 6,600 feet on the flat, undulating seafloor of a vent site radically different from the nine others they had seen since the expedition began. Guaymas Basin is covered with thick, organic-rich sediments from the outflow of Mexican rivers. Its multiple mineral mounds and upwelling lavas intrude as sills within this sediment. The combination of organic-rich sediments and mineral sulfides creates high concentrations of petroleum products. According to one scientist who has seen the site many times, when you dive Guaymas, you are touring one of Poseidon's petroleum refineries.

Guaymas is also known for its strange, pagoda-shaped mineral structures and thick microbial mats. Nearby are giant tube worms, scale worms, and polychaete worms. The sediments among the structures and worms are covered with dense layers of sulfide-oxidizing bacteria called *Beggiatoa*. In places the cream-colored bacteria build mats thick enough to scoop up with a spoon.

*Mats of white bacteria—critical to the food chain in a world without sun—metabolize hydrogen sulfide and in turn feed vent animals such as the crab above.*

They got their bearings and quickly found one of the big, flat-flanged mushroom structures rising out of the sediments. Its outcropping ledges were formed where high-temperature hydrothermal fluid escaped laterally. Underneath the ledges were wide pools of the buoyant fluid. Caught in the sub's lights, the silvery pools showed an infinite variety of reflections, and ripples moving in all directions.

They parked the sub in front of one of the wide brim structures, and under Conrad's direction deployed an instrument called McDUVE (multi-channel deep ultraviolet excitation). The device, which is basically an organic molecule prospector, was developed by the Jet Propulsion Laboratory in collaboration with Photon Systems of Covina, California. It exploits a phenomenon known as laser-induced native fluorescence, a characteristic of many organic molecules.

"It's basically a big ray gun that you aim at a target," says Conrad. "The illuminating beam—a

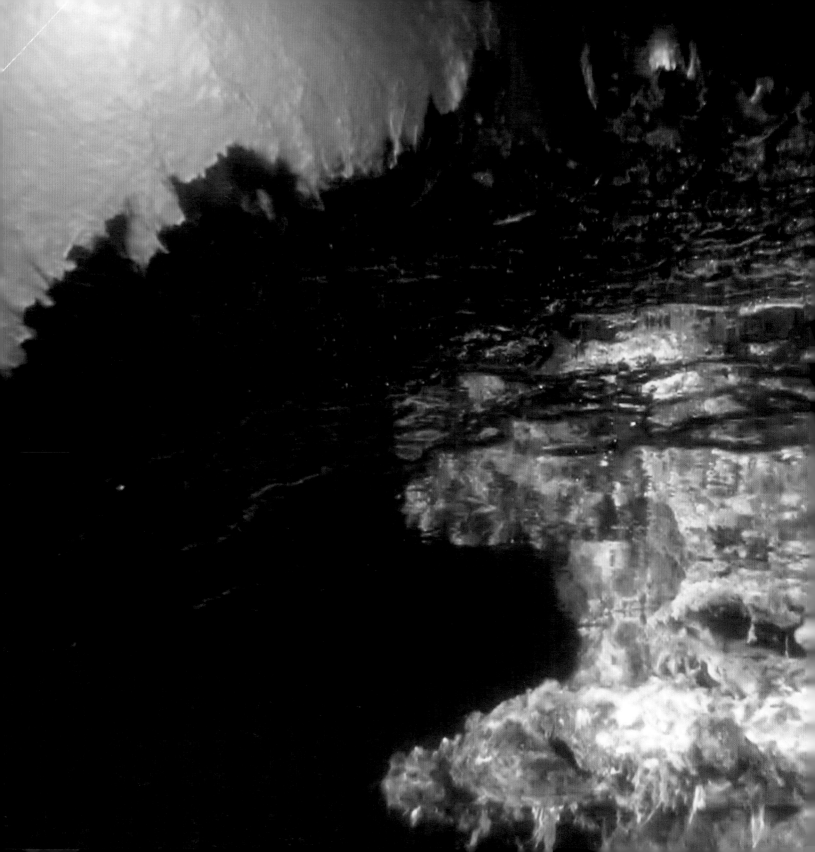

deep ultraviolet laser—is invisible, but if the target possess a variety of organic molecules and even some inorganic molecules, it will glow. The color of the glow reveals something about the type of the molecules. It doesn't conclusively identify life, but it does tell you that you are observing organic molecules."

While Cameron filmed McDUVE in action, Conrad adjusted its controls and logged the readings in her laptop computer. On this dive she mapped the distribution of organic molecules around one of the vent openings.

It is the second time the precision instrument has been used on the expedition. Next summer Conrad and her team are taking McDUVE to the high Arctic to see if they can tell the difference between rocks weathered in the presence of life and those weathered by abiological geochemical processes. They are also considering the possibility of a future deployment at the bottom of Lake Vostok in Antarctica. The huge subglacial lake, which lies under 13,000 feet of ice, has a thick sedimentary bed, and there is the remote possibility of venting structures within the sediments, and perhaps primitive forms of life. If the JPL team could successfully maneuver a hypersterilized instrument down through two miles of ice, they would be on their way to building a "cryobot" probe that could fly to Europa, tunnel through its thick ice cover, and examine those faraway sediments.

Occasionally, Conrad leans over and looks out of the five-inch view port. The thick white filamentous mats passing below the sub are a reminder of how important bacteria are in the making and sustaining of life on Earth. Scientists theorize that the first living organism to appear on the planet 3.5 billion years ago was a single bacterial cell—the ur-ancestor of all life. For the next two billion years, the longest stage in the evolution of the biosphere, the Earth was inhabited solely by microbes expanding and altering themselves and laying down the rules of existence on the third planet from the sun. They completely transformed the lithosphere and atmosphere. They designed the tens of thousands of miniaturized chemical systems

OPPOSITE: *Ripples of vent fluid stream out in all directions at a typical Guaymas Basin "mushroom," formed by the lateral path of hydrothermal fluid.*

essential to life, including photosynthesis and oxygen breathing.

For Conrad, the human body is a living history of life on Earth. Each of our trillions of cells contains water and salts similar to the composition of the primal seas. We are what we are because of bacterial partners working away in a watery environment.

From birth, the human body houses a Noah's ark of microbes working on our behalf. The bacterial colonies within us are among the densest in nature, exceeding the number of our body cells by more than tenfold. Not only do we co-exist with microbes, we are home to the remnants of some that have been subsumed inside our cells. Among them are descendants of bacteria that once floated in the early ocean and now exist in our brains as mitochondria. These tiny, membrane-wrapped inclusions enable us to use oxygen and help power our thinking. It is not inconceivable that the consciousness enabling us to explore the ocean may have arisen from countless microbes evolving symbiotically to shape the human brain.

When Conrad looks out the view port she looks into a world where nothing is permanent. Guaymas Basin is not one place but a thousand places, each with different ecological features. It is ruled by chaos and order, transition and mutation, where everything is exceptional and strange, and has been this way forever.

Cameron takes his eyes off the monitor and looks around the interior of *Mir One*. He loves this old sub with its curved walls and confined space, its dim lights and dials and its humming electronics. During the magic of a dive, the sub seems to be enriched and become a living presence. Every object within it glows with an animate vitality of its own—the oxygen monitor, the communication microphone, the blunt, curved joystick and wall-mounted witness camera recording his every move.

Cameron's been driving around in extreme-depth subs for almost ten years, but still loves everything about the experience: the going down through the long miles of water, the looking around in the immeasurable darkness, and the coming back up. With a sharp stab of recognition, he reminds himself that the expedition is almost over and this is the last time that he will be in the innermost depths for a while.

As *Mir One* approaches the surface, the three people inside see the first gleam of the sunlit water moving above them. "A short time ago," thinks Cameron, "we were deep inside the ocean, so deep that even without the water, we'd be hard to see. We were communicating, navigating, and working in a place of isolation and darkness. Now we're coming back up to Earth. It's like a space mission in reverse. When we recover, we actually splash up."

The moment the sub breaks the surface, Sagalevitch is on the intercom, giving directions to the small towboat coming out to fetch them.

Cameron never shoots the recoveries. For him, they are a waste of videotape because the routine is always the same: Wait until the towboat comes and attaches a line to the front of the sub; get towed until you are 50 feet away from the starboard side of the *Keldysh;* hold tight until the Russian "cowboys" arrive in their inflatable boat and one of them leaps onto the top of the *Mir* to secure the lift hook; wait until the big articulating crane

*JPL scientist Pan Conrad looks out of the small window of a* Mir *sub.*

reaches down and yanks the sub free of the sea and up onto the main deck. On a good day a recovery takes 30 minutes. On a bad day, when the swells are running and the sub is rolling like a greased barrel, it can take two hours. On a bad day, you close your eyes, curl up in the fetal position, and pray for deliverance.

But when Cameron glances at his monitor he sees the evening sunlight slanting into the water, backlighting a blizzard of plankton, transforming the ocean into the primordial sea of three billion years ago. The opportunity is too good to pass up. With a practiced motion, he pulls a tape out of the bag, unpops its tab, presses it back into the recorder, and starts shooting.

Cameron pans the camera around slowly until a large, dark object begins to fill his monitor. Its shape and its shadows are confusing until he realizes with a jolt that he is viewing the hull of the *Keldysh,* a 450-foot, 8,000-ton ship that must be right above them, and if it's right above them they're headed for a collision.

# ANATOMY OF A TUBE WORM

When the first pictures of hydrothermal vents were seen in 1977, the animals that grabbed everyone's attention were the tube worms. The white crabs and foot-long clams at the vents looked familiar, but the scarlet-tipped creatures sticking out of cream-colored tubes were startlingly novel. They were big. They had garish red tops and jabbed out of the bottom like a swarm of rigid snakes. They were not just a new species; they were a new phylum.

These worms, whose tubes can grow to a length of eight feet, have two striking anatomical features. They have no mouth or digestive system. And within their white tubes they have a specialized organ called the trophosome that houses sulfide-oxidizing bacteria. The tube worms could not exist without the presence of this bacteria.

Tube worms, or *Riftia pachyptilia*, have four distinct body regions: the tentacular, blood-red plume, a muscular collar or vestimentum, a circular trunk, and a short segmented base. The trunk contains the trophosome.

The worm lives inside the cylindrical tube made of a tough, white substance called chitin. The base of the chitin is usually anchored to rock. Under normal conditions, the worm ascends the tube to expose its plume to the surrounding water. If it senses the presence of a crab or other predator, it quickly withdraws back into the tube.

The plume is composed of tightly stacked sheets of finely divided tentacles. The sheets are supported by a central column and form a gill-like organ with a large surface area for the uptake of nutrients. A network of blood vessels within the plume allows efficient exchange of dissolved molecules between blood and water.

The worm has two primary blood vessels that circulate blood from the plume to the trophosome and back again. As its crimson color indicates, the blood is rich in extracellular hemoglobin.

The worm's muscular collar holds it in the tube and allows it to slide up and down. The heart and brain of the worm are housed in this region, as are its paired genital pores.

The trophosome tissue is soft, dark green in color, and laced with blood vessels. Internally, it is lined with host cells or bacteriocytes that are densely packed with bacteria. Each gram of trophosome tissue contains several trillion bacteria.

Tube worms huddle in clefts and hollows a safe distance from the vents, where they seek a good flow of dilute fluid at a maximum temperature of 59°F to 68°F. This allows them to simultaneously absorb oxygen from the cold seawater and hydrogen sulfide from the warm vent fluid. Both metabolites are absorbed by the plume and carried by the bloodstream to the trophosome. When oxidized by the symbiotic bacteria, hydrogen sulfide yields considerable energy.

Tube worms live in a tightly confined world and can't survive anywhere else. They conquer new territory by spawning enormous clouds of larvae into the water. The larvae drift in the currents until they sense the heat and chemicals of a welcoming hot spring. Then they descend and hunker down on its corrugated substrate. Like ancient seafarers, tube worms island-hop their way across entire oceans.

OPPOSITE: *With no mouth and no stomach, tube worms are among the most unusual animals found at hydrothermal vents.*

"Hey, Tolya," he says to Anatoly, "The hull is getting close. Do we normally get this close?"

Pan Conrad, who has seen a lot of strange sights in her life, is staring at the monitor with her eyes getting wider by the second. She feels like she's trapped inside some pulse-pumping video game.

"Tolya, look, look," says Cameron, his voice rising half a notch.

The intercom crackles with an urgent Russian voice from the towboat. Sagalevitch leans forward into the center view port, his hand tight on the joystick, and powers the sub away from the towering mass of steel above them.

A large nasty thump rocks the stern of the sub, followed by another. The third thump contains a sound like the squeal of a swamp pig.

Inside the sub, everything is silent except the hum of the carbon dioxide scrubber. Three people look at each other. Each of them waits for the thunderclap that never comes.

High above, looking down from the railing of the *Keldysh,* a line of spectators sees a piece of white fiberglass floating away from the sub's faring and the broken starboard light boom dangling in the water.

Conrad smiles and gives Cameron one of those "old-hands-at-sea" looks. Not ten minutes earlier they had been talking about how well the dive had gone, and Cameron said when things are going really well is when you should start to worry. "The moment you think what you are doing is routine," he said, "that's when things go sideways."

A few minutes later, *Mir One* is lifted slowly out of the Gulf of California carrying three people who lead regular lives living in cities. But on certain days they climb aboard a ship, sail across an ocean, dive into its heart, and glimpse things never before imagined.

The man in the blue jumpsuit crouched next to the video monitor is looking out the view port at the unsettled surface of the ocean. James Cameron has spent a lifetime exploring the outer edge of human possibility and is already thinking about his next undersea project. For James Cameron, there is always another horizon, another dream.

OPPOSITE: *Nicknamed "Dumbo" octopus, this benthic cephalopod uses its fins to help it swim.*

The spectacular geo-
graphy of 9° North was
created by lava, explains
Maya Tolstoy. At one
point lava flowed as
high as the top of these
bridge-like structures,
but layers broke away
as the lava cooled. The
structure that is left
represents cooler,
denser lava.

This benthic "thorny-head" sports a smart color for a world that is lit only by occasional bioluminescence: That light, which is mostly blue, is absorbed by the red, keeping the fish invisible in the dark abyss.

**ABOVE:** A vent crab catches an unexpected ride on a sample gathered from the arm of the *Deep Rover* sub. **RIGHT:** Lights from a *Mir* illuminate an unusual

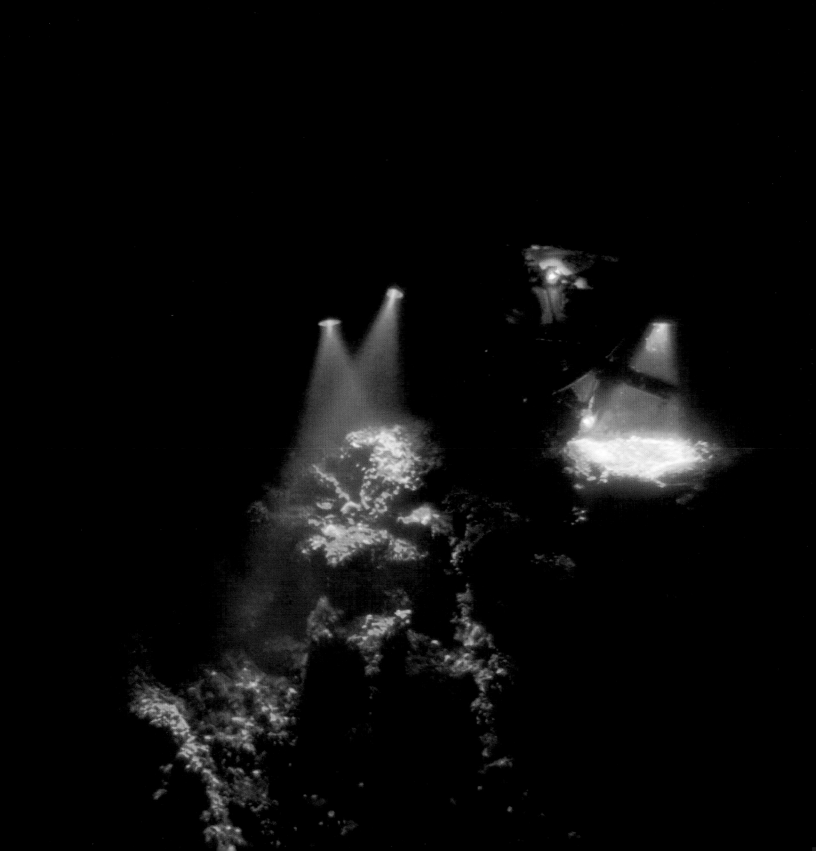

The filter-feeding
Brisingid starfish at
right catches passing
plankton and detritus
with its long arms, then
passes the food to its
mouth using special
gutters that run the
length of each arm.

Formed by the
repeated heating and
cooling of lava under-
water, a bed of pillow
lava blankets the
seafloor around the
vent at 21°North.

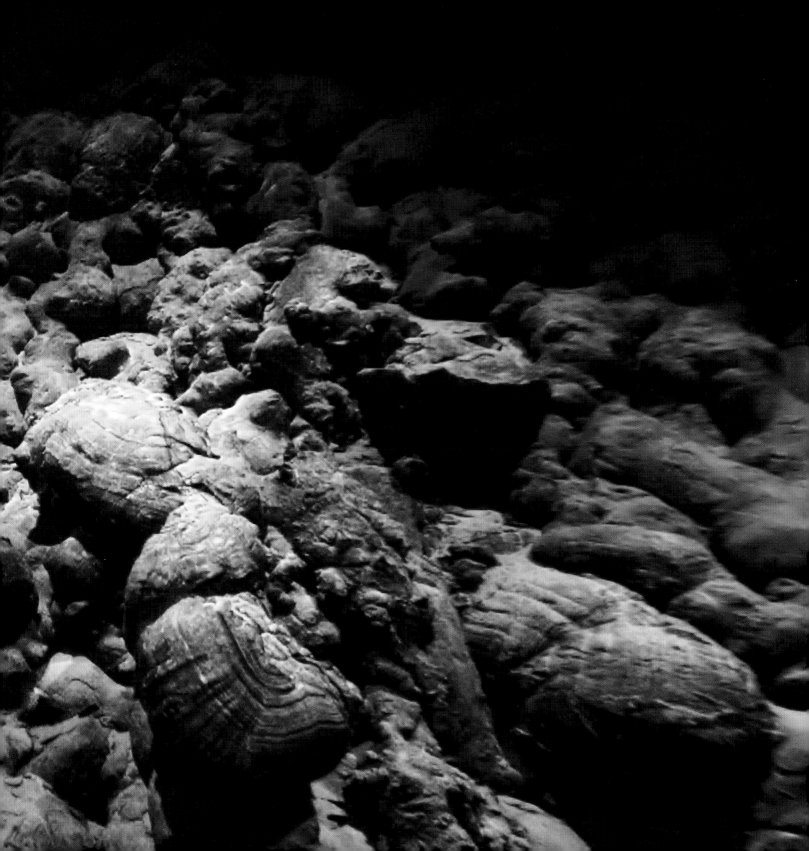

A vent octopus is
captured in a ghostly
dance at 9°North.

160

Indicators of a vent site in decline, vent crabs scavenge a ravaged field of empty tube worm stalks. Death can come in many forms at a vent: Volcanic eruptions can decimate an ecosystem quickly, while the slow, steady movement of the Earth's crust may gradually deplete a vent community of the hydrothermal fluid that fuels its existence.

While the tubes of healthy tube worms can grow to lengths in excess of 10 feet, the bodies of the worms inside the tubes rarely exceed four feet. Within a few years of birth, the worms are fully grown.

# CHAPTER FIVE

# MAKING THE FILM

## FIRST CONCEPT TO FINAL CUT

One of the expedition's four $250,000 3-D-HD cameras is tilted down for a close-up.

From the beginning, Jim Cameron knew that making the film was going to be a formidable task. A compelling film treatment had to be written. Millions of dollars had to be raised from a studio. A first-class documentary film crew had to be selected. The two *Deep Rover* subs had to be completely rebuilt. Four 3-D-HD cameras had to be checked and tested, and two of them had to be installed in titanium housings and mounted on the bow

*In the presence of cameras and microphones, Loretta Hidalgo prepares for her first dive.*

of *Mir One* and *Rover 1*. Thousands of pieces of equipment, including standard video cameras, cables, camera cranes, recording tapes, editing equipment, and all their spare parts, had to work for months in high winds and waves, in blistering summer heat, and—when the subs were submerged—in a corrosive, near-freezing, high-pressure environment. Afterward, when the ships sailed into port, miles of footage would have to be edited and merged with sound and special effects.

All major films have three phases. The first is preproduction, where the treatment or outline is written, the budget is estimated, the equipment is prepared, the crew is selected, and schedules are arranged. The second is production, where the film is shot, reshot, recorded, and studied. The final phase is postproduction, where the footage is edited, the sound track is designed, and special effects are added to make a dramatic visual statement.

Under the direction of Andrew Wight, the producer, and Ed Marsh, the creative producer and film editor, a timeline was developed. For *Aliens of the Deep*, preproduction would take place between February and June 2003. Production would occur between July and October. In November, Ed Marsh and his co-editors Fiona Wight and Matt Kregor would begin to edit together the first chapters. The finished 45-minute film would be delivered the following autumn and released in January 2005.

OPPOSITE: *Jim Cameron films the* Deep Rover *sub team preparing* Rover 2 *for a dive.*

The first step in postproduction is to log all the scenes shot on the surface and under the sea. A typical log list reads like the one from Dive 23 in the Pacific at the Guaymas vent site. *Scene 1: A white octopus sitting on the bottom. Scene 2: The JPL instrument points at bacteria. Scene 3:* Mir One's *manipulator takes a sample of the bacteria.* There are 25 scenes on this 60-minute 3-D-HD roll—one of 350 rolls shot during the expedition. That's not counting more than a thousand hours of standard video. The almost 1,500 hours of video explains why it took weeks to log the footage.

In November 2003, production assistants began working with Ed Marsh to log all the tapes and identify the most impressive images. Then they began to edit together the first "chapters within the story" that would become the rough assembly. Looking at the moving pictures over and over, searching for the best way to link them together, they saw images that amazed them. Scientists jammed into the *Rover* with the floor space like a narrow steel coffin and the head space claustrophobic. The *Rover* pilots, sweat shining on their faces like olive oil, their T-shirts clinging to their shoulders like sodden rags. Golden beams of light jabbing through the black water and lighting up another sub. Carbonate chimneys as high as six-story buildings standing unbroken and impregnable. Seething walls of shrimp, every head and torso and leg wrapped in a vesture of heat.

They examined the images so long they began to imagine the rumble of fluid coming out of the vents, a sound like the grinding of Earth's great geological engine.

At the Earthship offices, editors worked into the night. Every day, they passed a chart on the wall that reminded them of approaching deadlines. *December 1: Begin rough assembly edit. January 12: Vendors deliver temporary visual effects. January 26: Complete assembly for Jim's assessment. February 23: Second pass at the edit.*

In April, after the pickup shots, green-screen shots, and interviews had been added, the rough assembly was about 60 minutes long. By June it had been cut to 47 minutes, and Jim Cameron came in to review it, for the fifth time. The following excerpts from his 13 pages of suggestions show his attention to detail.

OPPOSITE: *Sunset brings a lighting challenge on the* Keldysh. *For three months the film crew worked around the clock.*

## *ALIENS OF THE DEEP* CUT 5 NOTES

01:18 ◀ *The CG (computer graphic) sun shot is too short. Needs to be 25% longer to give it more power. For maximum 3-D effect, we should be close enough so that a solar flare is roaring right out at us.*

10:16 ◀ *For this dive at Lost City and the next one, we might start the sequence with a super which says LOST CITY and the date and time. This will make it seem more technical, like a real expedition. We could also do it with Maya's dive, and other places where it is appropriate.*

20:00 ◀ *I shot some stuff of Maya concerned they were flying into a black smoker. We need to feel the jeopardy here…words about the heat of the smoke and that it can melt through the sub's faring…I want to see a proposal for a sequence of shots that create a sense of hazard…*

21:03 ◀ *These great shots of the Crown of Fire at Lucky Strike don't work here. The value of these shots is to show how the animals have adapted to living next to what is basically open flame. They're right next to the erupting fluid, tickling the dragon's tail. This concept comes later so the shot should come later. We've got some strong shots of powerful black smokers. They need to go here.*

*We should forget about the bacteria here and show the other forms of life. The sequence should go:*

◀ We're approaching some chimneys.

*Reveal. Wow…. Chimneys are cool.*

*Chimneys are big, loud, violent, turbulent underwater volcanoes (music, ominous, powerful)*

*Jeopardy. We're in the smoke…arrrgh we're dying (not really, but we could have)*

*And WOW!! Chimneys support life. Creatures live near this stuff.*

▲ *And somehow these critters have adapted to the heat. They taunt death with every move, swimming and crawling right near the super hot water (crabs in shimmering water, shrimp in the smoke) The stuff in the black smoke is keeping them alive.*
*And WOW!!! There's an insane amount of biomass. This place is teeming with life…an insane profusion (use the boiling shrimp in the Snake Pit shots)…. Dante's inferno meets Chinese rush hour. Now we go to the swing idea: wait a minute… if we don't need the sun for life…for robust, energetic life…and all we need is water and some kind of energy source… there is another place in the solar system that has THAT! Dissolve to EUROPA.*

In September 2004, the final visual effects were integrated into the film and the final sound design was laid down. In mid-October, the finished film, called the negative, was delivered.

# SHOOTING IN 3-D

Because our eyes are about three inches apart, we see the same object from two slightly different angles. This binocular or 3-D vision allows us to see things more clearly and better determine their distance and movement.

Cameras that film in 3-D apply this same concept to capturing moving images. Working with engineers from Sony and Panavision, Jim Cameron and Vince Pace applied this concept in the design of their unique 3-D camera system by modifying two Sony 950 high-definition video cameras and placing them side by side—as close together as a pair of human eyes—in a special case.

Capturing 3-D action on the deck of a rolling ship requires a closely synchronized team: a cameraman to film with the 30-pound camera package; a stenographer to control the angle of focus, zoom, and convergence, and the spacing between the lenses; a focus controller to adjust the camera's focus; and a fourth person to carry the battery and electronics package.

To record the action deep under the sea, components of the Sony cameras are placed in a pressure-protective titanium housing built by Mike Cameron and installed outside the pressure hull of *Mir One* and *Rover 1*. The housing is the largest implodable volume ever attached to a manned sub. If it fails at 12,000 feet, the force of the implosion will almost certainly destroy the sub and everyone inside. "It's like living with high explosives strapped to your pressure hull," says Vince Pace.

During a dive, the cameraman sits beside the pilot looking at the camera's viewfinders and a 9-inch HD monitor. The sub is neutrally buoyant and influenced by currents, so the pilot is crucial to the success of almost every shot. The images are sent through cables and high-density pin connectors to digital recording decks inside the sub's pressure sphere.

"Shooting 3-D high-definition cameras on a heaving deck or from inside a sub at 8,000 feet is the easy part," says Pace. "The hard part is making sure that the images are being recorded exactly as you shoot them. The subs get banged around a lot. Salt eats through wires and connectors. You've got thousands of complex parts that all have to work together all at the same time. The most important component is the team that keeps them working."

John Brooks has been a cameraman and documentary director for more than 30 years. "I've never been more challenged than on the *Aliens* film shoot," he says. "Eight of us worked on every deck on both ships, shooting day and night.

"One day we were on the deck of the *Keldysh* after a long dive. Dozens of people were emptying the sample baskets in front of the subs and swarming around to see what was in them. Jim, who had just spent nine hours inside *Mir One*, grabbed a 3-D camera and started shooting. Although the camera weighs about 30 pounds, he moved through the crowd with the shutter wide open, moving in close, stepping back, and then moving in close again. He was creating visual choreography out of visual chaos, and in his hands the camera seemed weightless. There were dozens of places he could have taken his camera, but he's got a sixth sense for the one that best drives the story."

OPPOSITE: *A 3-D-HD camera in its scuba housing.*

As a young man, Jim Cameron's passion for science and technology led to a personal interest in exploring the limits of the known world. His curiosity about boundaries and what's beyond them influences everything he does. It's one of the themes that runs through his feature films, including *The Terminator* (1984), *Aliens* (1986), *The Abyss* (1989) and *Solaris* (2002), and it's the primary theme of his ocean expeditions and documentaries. It prompts him to push the invention of new technologies, from underwater lights that sweep away the darkness to special effects that transport our imaginations. On a personal level, it is one of the reasons he applies such furious energy to all his projects.

The ocean is a "frontier" in many of Cameron's films. In 1988, he produced and directed *The Abyss,* an action-adventure story about an offshore oil drilling crew recruited by the U.S. Navy to salvage a downed nuclear sub. Combining brilliant underwater photography with astonishing special effects, *The Abyss* had an extreme-depth undersea station, two transparent mini-subs, and a diver breathing an oxygenated fluid.

The film was the outcome of a concept Cameron had been thinking about since high school. The original story was set in a science station on the edge of a steep undersea canyon. "It was really about human curiosity," he says. "Highly trained divers wanting to know 'what's down there.' They go into the abyss and are lost. In the end the station is in ruins and there's only one guy left. Although he's certain he's not going to come back, he dives down to see what happened to the others. The story ends with him going deeper and deeper, urged on by his need to know."

Most of *The Abyss* was shot underwater in the containment vessel of an abandoned nuclear power plant in South Carolina. Working at a blistering pace over a period of four months, Cameron, the actors, and the film crew made more than 10,000 dives on the submerged set. It has been called "the hardest film shoot in Hollywood history."

Cameron's goal of filming *The Abyss* underwater led to the development of first-ever lighting systems, camera gear, propulsion vehicles, and a subsurface communication hook-up that allowed him helmet-to-helmet contact with his cast and crew.

*The Abyss* also introduced some astonishing computer-generated images in the form of a gleaming, watery alien. The transparent tentacle that could mimic human faces was so authentic that it won an Academy Award for special effects supervisor John Bruno.

Cameron's interest in *Titanic* began with the discovery of the wreck by Robert Ballard in 1985. He wanted to make a film that integrated what had been learned about the *Titanic* since its discovery with the events of 1912. "My goal," Cameron said, "was to will the *Titanic* back to life."

He began by chartering the *Keldysh* and making 12 *Mir* dives to film the wreck. "On my second dive it suddenly hit me," he says. "I was on the *Titanic*. Over eight decades dissolved in an instant. I was walking on the deck with those soon-to-be-famous—or infamous—passengers: J. Bruce Ismay, Captain Edward J. Smith, Thomas Andrews, and the 1500 souls who perished when the great liner sank."

*The massive prow of* Titanic *is still intact after more than 90 years on the ocean floor.*

After the dive, back on board the *Keldysh*, Cameron was deeply moved by what he had seen and felt. He realized that his film was doomed to failure unless it conveyed the emotion of the night as well as the fact of it. The experience led him to create the characters of Jack and Rose, the fictional couple whose love story humanizes the tragedy. "It draws you in," he says, "and allows you to visit the most interesting parts of the ship. It is what brings the movie full circle: from a film about the *Titanic*, to a love story set on the ship, back to the truth about the *Titanic*."

The film took three years to make and rewrote motion picture history. Like *The Abyss*, it involved the development of a suite of new technologies from camera booms to underwater lights to underwater housings. Its special effects, from the great ship's departure in Southampton to her sinking into the swirling Atlantic, were breathtaking. *Titanic* won 11 Academy Awards, including Best Picture and Best Director.

*Lights from* Mir Two *cast a ghostly glow on* Titanic's *portside Boat Deck near the officer's quarters and Marconi room.*

After the stunning success of *Titanic,* Cameron did what he always does after a brutal film shoot: He went on a diving trip. "Filmmaking is insanity," he says. "You take the phone off the hook and you don't answer it for a year. The ocean is where I go to regain my sanity." This time, instead of going to the western Pacific or the Caribbean to scuba dive, he decided to have a look at the real abyss. His dives to *Titanic* confirmed that the *Mir*

vehicles were a pair of reliable shuttles to inner space. The challenge was to devise some meaningful deep-sea work and a way to pay for it.

His solution was to make another film about the ship he'd won his Oscars for. "I began thinking about what it would be like to explore her interior with a pair of mini-robots. And use this vantage point to tell the stories of the real people—some who died, and some who lived—

on the ship," he said. "Because *Titanic* is overgrown with rusticles and is badly damaged, we would use computer animation to show it the way it was and then come back to present-day. To make the story more real, we would film everything in three dimensions."

Released in 2003, *Ghosts of the Abyss* was the first 3-D film ever shot in the deep ocean. In it, the actor Bill Paxton and a group of *Titanic* experts, including a biologist, an artist, and a historian, went down to take their first look at the ship that was slowly dissolving into the ocean. "Our objective," says Cameron, "was to make the audience feel they were passengers on the ship. The only way to do that was to shoot it in 3-D."

For years Cameron has been advocating the development of a rugged, flexible 3-D "reality-camera" both for feature and documentary films. One of the places he'd like to take it is space. He underwent training in Russia to prepare for a trip to the then-orbiting *Mir* space station. He talked to NASA about joining the astronauts on a mission so he could film in orbit. For Cameron the two most exciting words are "terra incognita"—unknown territory. The next five are, "Yeah, we could film that."

Cameron's fascination with alien landscapes focuses on shipwrecks like *Titanic* and *Bismarck* because they embody timeless places and enduring myths. "The German battleship *Bismarck* was the Death Star of its day," he says. "I wanted to tell her story because this indomitable killing machine from World War II contained so many echoes of what's happening in the world today."

Adolf Hitler's navy commissioned the 42,000-ton warship in 1940. Its attacks on the convoys sailing from America to Britain with food and war matériel were devastating. In 1941, the *Bismarck* sank Britain's great battle cruiser the *Hood*, taking all but 3 of her 1,400 men down with her. Not long after, a British armada found the *Bismarck* and bombarded her with shells and torpedoes. Only 116 of her 2,200 sailors survived.

*James Cameron's Expedition* Bismarck premiered in 2002. One of its highlights was the mini-robots that allowed the audience to "fly" through the interior of the ship lying 16,000 feet under the Atlantic. The unprecedented images were combined with computer animation, archival film, dramatic reenactments, and interviews with surviving seamen and experts on the sinking.

James Cameron is a 21st-century explorer, equally at home riding a sub to 16,000 feet to film the *Bismarck* as he is sitting alone in a room editing the final version of *Aliens of the Deep*. For him, life is about exploration and exploration stories that leave the world a little wiser.

"Our culture is getting apathetic about exploration," he says. "A lot of people feel the whole world is mapped and it's fine for robots to go out and examine Mars, but human exploration is all done. But that's absolutely not the case. We've barely scratched the surface in space or under the ocean. Sign me up. I'll go either way."

I am standing at the edge of the ocean near the Earthship production offices in Malibu, California. It is a warm spring day, and the late afternoon sun flashes off the tops of the big Pacific swells as they break into foam and flatten against the beach. From where I stand, the floor of the ocean slopes steadily downward into a place that has never known sunlight. Its hidden sand and sediments run out to the horizon, and out to the next horizon and the next. I am perched on the edge of the greatest wilderness on the planet.

I look out at our home star as it eases into the Pacific. For the past year and a half, I have been part of a team that made extensive preparations and then traveled thousands of miles across the Atlantic and Pacific Oceans on a mission of cinematic discovery. I have had the good fortune to see the most advanced filming and submarine technologies deployed deep within the sea at places with names like Snake Pit and Lost City. For the first time, I learned the real meaning of the Mid-Oceanic Ridge, volcanic vents, microbes, and Mars. I saw how four hundred hours of 3-D high-definition footage is patiently edited into a compelling 45-minute large format film. Most important, I had the opportunity to go to sea with some of the finest, hardest-working, most resilient men and women I have ever met. We were led by a man who is changing the way the world sees the ocean and its relationship to life in other parts of the solar system. The future may show that unlocking the secrets of deep-sea life is an essential step in the discovery of alien species in what Jim Cameron calls "the other great blackness."

*James Cameron and Sebastian Vega chase the perfect shot.*

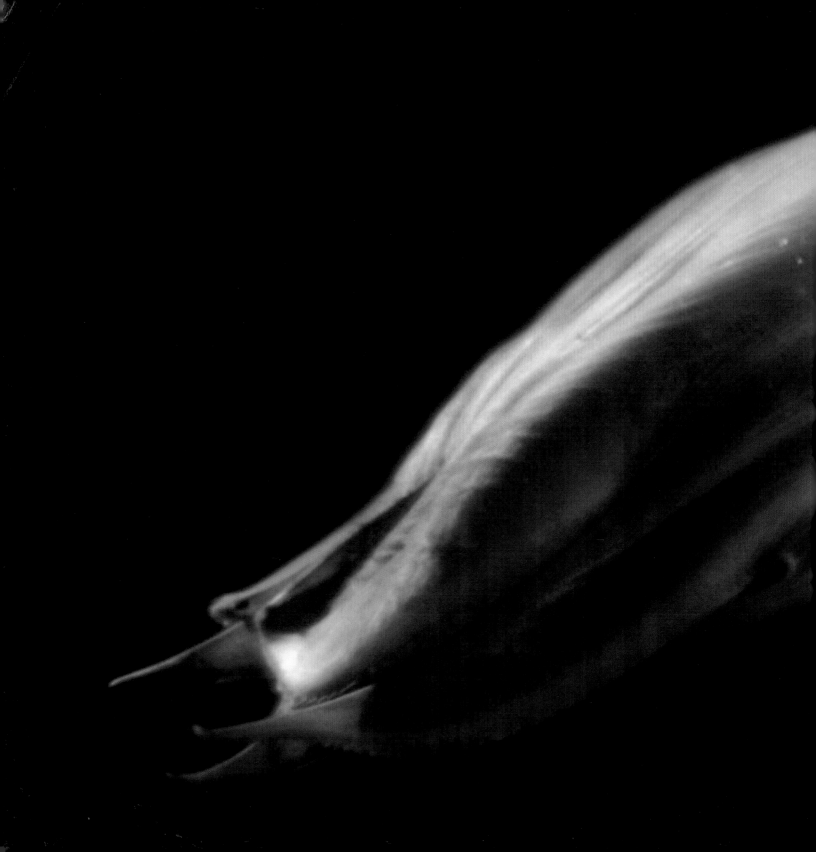

An octopus stops to
investigate the strange
intruders in its under-
sea home.

Achenbach, Joel. *Captured by Aliens: The Search for Life and Truth in a Very Large Universe.* Simon and Schuster, 1999.

Ballard, Robert D., *Exploring Our Living Planet.* National Geographic Society, 1983.

Dick, Steven J. *The Biological Universe: The Twentieth Century Extraterrestrial Debate and the Limits of Science.* Cambridge University Press, 1996.

Edmond, John M. and Karen Von Dam. "Hot Springs on the Ocean Floor." *Scientific American,* April 1983: 78-93.

Ellis, Richard. *Deep Atlantic: Life, Death, and Exploration in the Abyss.* Alfred A. Knopf, New York, 1996.

Grinspoon, David. *Lonely Planets: The Natural Philosophy of Alien Life.* Ecco/HarperCollins, New York, 2003.

Hartmann, William K. and Ron Miller. *The Grand Tour: A Traveller's Guide to the Solar System.* Workman Publishing, New York, 1991.

Kaharl, Victoria A. *Water Baby: The Story of Alvin.* New York, Oxford University Press, 1990.

Kelly, Deborah et al. "An off-axis hydrothermal vent field near the Mid-Atlantic Ridge at 30ºN." *Nature,* Vol 412, July 12, 2001.

Kunzig, Robert. *Mapping the Deep: The Extraordinary Story of Ocean Science.* W.W. Norton, New York, 2000.

Lutz, Richard A. and Rachel M. Haymon. "Rebirth of a Deep-sea Vent." NATIONAL GEOGRAPHIC, Vol. 186 (November, 1994):114-126.

MacDonald, Kenneth C. and Paul J.Fox. "The Mid-Ocean Ridge." *Scientific American,* June 1990:72-79.

MacInnis, Joseph. *Breathing Underwater: The Quest to Live in the Sea.* Penguin Canada, Toronto 2004.

Margulis, Lynn and Sagan, Dorion. *Microcosmos: Four Billion Years of Microbial Evolution.* Simon and Schuster, New York, 1986.

Morton, Oliver. *Mapping Mars: Science, Imagination and the Birth of a World.* Picador, New York, 2002.

*Oceanus,* Vol. 34 (1991) Number 4: Mid-Ocean Ridges.

Sagan, Carl. *Cosmos.* Random House, New York, 1980.

Smil, Vaclav. *Energies: An Illustrated Guide to the Biosphere and Civilization.* The MIT Press, Cambridge, Mass, 1999

Van Dover, Cindy Lee. *The Ecology of Deep-Sea Hydrothermal Vents.* Princeton University Press, Princeton, N.J., 2000.

Van Dover, Cindy Lee. *The Octopus's Garden: Hydrothermal Vents and Other Mysteries of the Deep Sea.* Addison-Wellsley, Reading, Massachusetts, 1996.

First and foremost, my deepest thanks to Jim Cameron. His passion for exploration inspired every word and image in this book.

To Andrew Wight, executive producer, for his cool camera, cool assessments, and chilling cave-diving stories.

To Ed Marsh, the film's creative producer, John C. Anderson, and Lucas Sanders, who brought the high-definition images from the giant screen to the pages of this book. To Steve Quale for his insight on how the film was made, and to Fiona Wight and Justin Shaw for their assistance.

To all those on the good ship *Ares,* especially the *Deep Rover* sub crew—Patrick Lahey, Dave Norquist, Tym Catterson, Paul McAfferty, Tim Bulman, Ed Hoeffing, Dano Pagenkopf and Kris Newman—for the long weeks of advice, humor, and insight, and the brief sub ride to the outskirts of Lost City. No one could ask for better shipmates. And to Captain Dennis for his astonishing ship-handling abilities and his Greek music and coffee.

To Vince Pace and his camera team, including John Brooks, Jeff Cree, John Malvino and Mark Robinson for showing me how they keep the gear running and make a 3-D-HD film when the temperature is 90 degrees in the shade.

To J .D. Cameron for keeping me safe on deck, safe on the water, and for sharing his Marine combat stories. And Mike Cameron for showing me what it takes to build the world's most advanced mini-robot, give it sea legs and fly it into the mouth of a volcanic vent.

To all the scientists, including Megan McArthur, Michael Atkins, Pan Conrad, Jim Childress, Michael Eastwood, Dijanna Figueroa, Kevin Hand, Tori Hoehler Loretta Hildalgo, Lonne Lane, Christy Reed, Kelly Snook, and Maya Tolstoy, for introducing me to their fascinating worlds of marine biology, marine geology, astrobiology, and planetary science. Every kid should have such enthusiastic and knowledgeable teachers.

To Debbie Kovacs at Walden Media, who steered this book through some shoal and stormy waters.

To Lisa Thomas at National Geographic Books for her early and sustained enthusiasm. And John Paine, for his editing skills.

And to Mike McDowell, Belinda Sawyer, Ronnie Allum, Chris Debeic, Atil Singh, Ralph White, Glen Singleman, Todd Cogan, Kristin Berbae, Melanie Miller, Ralph Burris, Rich Robles, Charlie Pelligrino, Lewis Abernathy, Daniel Greenwald, and all the others who were generous with their time and their talent—thanks.

—*Dr. Joseph MacInnis*

To go to sea, carry out the dives and produce the film, James Cameron assembled a team of some fifty men and women including Americans, Russians and Australians. Among them were scientists, sailors, sub pilots, technicians, production coordinators, cameramen and editors. They spent months on shore preparing for the expedition and months at sea on the *Ares* and *Keldysh*. Once the film was shot, they spent more than a year working in editing studios and sound mixing labs.

They were bound together by their professional skills and their commitment to this one-of-a-kind project and the passionate man in charge of it. Each of them made a unique contribution to the expedition and the film. Here are their names from the film's end credits:

Andrew Wight, Steven Quale, Ed Marsh, Jeehun Hwang, Vince Pace, Christopher A. Debiec, Chuck Cominsky, Bernie Laramie, Ron Allum, Ellie Smith, Ralph Burris, Atil Singh, Lilia Soto Aragon, Ralph White, Haley Jackson, Melanie Miller, Kristin Berbae, Wendy Lowe, Kendrick Hudson, Pete Hoffman, Leonard Barrit, Fiona Wight, Matt Kregor, Ed March, Justin Shaw, Todd Cogan, Neil Jariwala, John Brooks, Roy Unger, Dale Hunter, Mark Robinson, John Malvino, Sebastian Vega, Vincent Mata, Jeff Cree, Joe Vasquez, Glen Singleman, Mark Goodwin, Dwight Campbell, George Kallimanis, Tim Murphy, Denis Baxter, Daniel Greenwald, Adrian Degroot, Geoff Burdick, Joe Hagg, Al Rives, Lewis Abernathy, Mike Cameron/Dark Matter LLC, Peter Mortenson, Donald Keith Burgess, Brian Westfall, John David Cameron, Rich Robles, Patrick Lahey, Dave Norquist, Tym Catterson, Tim Bullman, Paul McAfferty, Ed Hoefing, Dano Pagenkopf, Mike Camp, Kris Newman, Christina Reed, Pamela "Pan" Conrad, Arthur "Lonne" Lane, Jim Childress, Dijanna Figueroa, Michael Henry, Maya Tolstoy, David L. Dubois, Michael Atkins, Kelly Snook , Megan McArthur, Tori Hoehler, Kevin Peter Hand, Loretta Hidalgo, Charles Pelligrino, Joe MacInnis, Ricardo Santos, Alberto "Al" Lopez, Michael Enriquez, Claudia Huerta.

With special thanks to the crew of *R/V Akademik Keldysh* and the P.P. Shirshov Institute of Oceanology of the Russian Academy of Sciences, Mike McDowell, Belinda Sawyer, Peter McDowell, Michael Eastwood, Darios Melas, Captain Dennis and the crew of the *EDT Ares*, Genya Chernaiev, Victor Nischeta, Anatoly Sagalevitch.

# ILLUSTRATIONS

All images from (c) 2004 Buena Vista Pictures Distribution and Walden Media. Walden Media is a registered trademark of Walden Media, LLC. All Rights Reserved – with the exception of the following: Blur Studios: page 172 up, CORBIS; page 26, Leonard de Selva/CORBIS; page30 CORBIS; page 34 Clouds Hill Imaging Ltd/CORBIS; page 92 Ralph White/CORBIS; page 93 Ralph White/CORBIS; page 95 CORBIS; page 96 Roger Ressmeyer/CORBIS; page 98 Reuters/NASA/ JPL/DLR/German Aerospace Center/CORBIS; page 106 NASA/JPL/Cornell/ZUMA/ CORBIS; page 110 Reuters/CORBIS; page 115 ESA/NASA and Albert Zijlstra/Reuters/CORBIS ; Creative Logik Universe: pages 120-1, 121 (both); David Doubilet: page 102; Drs. Michael J. Daly and Alexander I. Vasilenko, Department of Pathology, Uniformed Services University of the Health Sciences, Bethesda, MD 20814, USA.: page 105; Getty Images: page 12-13 Stone/Getty Images; NG Maps: pages 28, 29, 58, 136; Walden Media: pages 177 & 178, (c) 2003 by Walden Media, LLC. Used by permission of Walden Media, LLC from the GHOSTS OF THE ABYSS 3-D movie.

The Book Division is grateful to many individuals who contributed to *Aliens of the Deep*. Our special thanks goes to: James Cameron, Pamela "Pan" Conrad, Tori Hoehler, Jim Childress, Lucas Sanders, Blur Studios, Creative Logic Universe, Ed Marsh, Debbie Kovacs, John Paine, Judy Klein, Margot Browning, Emily McCarthy, Mike Cameron, Dark Matter LLC, and Brian Westfall. Brian's talent, good nature, and kind spirit is deeply missed.

# ALIENS OF THE DEEP

**PUBLISHED BY THE NATIONAL GEOGRAPHIC SOCIETY**

| | |
|---|---|
| John M. Fahey, Jr. | *President and Chief Executive Officer* |
| Gilbert M. Grosvenor | *Chairman of the Board* |
| Nina D. Hoffman | *Executive Vice President* |

**PREPARED BY THE BOOK DIVISION**

| | |
|---|---|
| Kevin Mulroy | *Vice President and Editor-in-Chief* |
| Charles Kogod | *Illustrations Director* |
| Marianne R. Koszorus | *Design Director* |

**STAFF FOR THIS BOOK**

| | |
|---|---|
| Lisa Thomas | *Project Editor* |
| John Paine | *Text Editor* |
| John C. Anderson | *Illustrations Editor* |
| Melissa Farris | *Art Director* |
| Lucas Sanders | *Image Enhancement* |
| Daniel O'Toole | *Researcher* |
| Carl Mehler | *Director of Maps* |
| Ed Marsh | *Picture Legends Writer* |
| Gary Colbert | *Production Director* |
| Rick Wain | *Production Project Manager* |
| Sharon Berry | *Illustrations Assistant* |
| Creative Logic Universe | *Computer Graphics* |
| Blur Studios | *Computer Graphics* |

**MANUFACTURING AND QUALITY CONTROL**

| | |
|---|---|
| Christopher A. Liedel | *Chief Financial Officer* |
| Phillip L. Schlosser | *Managing Director* |
| John T. Dunn | *Technical Director* |
| Vincent P. Ryan | *Manager* |
| Clifton M. Brown | *Manager* |

One of the world's largest nonprofit scientific and educational organizations, the National Geographic Society was founded in 1888 "for the increase and diffusion of geographic knowledge." Fulfilling this mission, the Society educates and inspires millions every day through its magazines, books, television programs, videos, maps and atlases, research grants, the National Geographic Bee, teacher workshops, and innovative classroom materials. The Society is supported through membership dues, charitable gifts, and income from the sale of its educational products. This support is vital to National Geographic's mission to increase global understanding and promote conservation of our planet through exploration, research, and education.

For more information, please call 1-800-NGS LINE (647-5463) or write to the following address:
National Geographic Society
1145 17th Street N.W.
Washington, D.C. 20036-4688 U.S.A.

Visit the Society's Web site at www.nationalgeographic.com.

Library of Congress Cataloging-in-Publication Data

MacInnis, Joseph B.
James Cameron's Aliens of the Deep
Joseph MacInnis; Introduction by James Cameron
p.cm.
Companion to the film Aliens of the Deep.
Includes bibliographical references and index.
ISBN 0-7922-9343-6

1. Hydrothermal vent animals. 2. Hydrothermal vents. 3. Aliens of the deep. I. Cameron, James 1954- II. Aliens of the deep. III. Title.

QL125.6.M325 2005
578.77'99—dc22

2004058319